MALIGNANT PIED PIPERS
OF OUR TIME

A Psychological Study of Destructive Cult Leaders from Rev. Jim Jones to Osama bin Laden

Peter A. Olsson, M.D.

PublishAmerica
Baltimore

First printing

At the specific preference of the author, PublishAmerica allowed
this work to remain exactly as the author intended, verbatim, without
editorial input.

ISBN: 1-4137-7668-X
PUBLISHED BY PUBLISHAMERICA, LLLP
www.publishamerica.com
Baltimore

Printed in the United States of America

This book is dedicated to my family and friends. My mother, Doris Olsson, had a deep faith in God and a joy in her friendships that inspires me to this day. I wish she could have lived long enough to read my book. My father, John Olsson, worked long hours to provide help for me in my life and my education. I miss them both.

My dear friends Bob White, M.D., Jules Bohnn, M.D., and John Newman of Houston have persistently provided me with constructive criticism and encouragement in my efforts to produce this book and my writing in general.

My daughter, Shannon Sweet, and my son Andy Olsson have helped me through their suggestions and enthusiasm. My son Nathaniel Olsson was very helpful at all phases of my work on this book. My wife, Pam Olsson, M.D., has contributed during many phases of the work that led to this book. Pam is the only tender guru I love and respect.

ACKNOWLEDGMENTS

I must acknowledge the courage and strength of Tim Stoen and Grace Stoen, who tried to free their son, John, from the web of Jim Jones's cult when they realized where the Malignant Pied Piper was heading. The Stoens' dilemma stirred my desire to understand and research the topic of destructive cults.

My mentor and friend Vamik Volkan, M.D., has encouraged and persistently supported my research efforts in this area. My teacher, C. Glen Cambor, M.D., at the Houston Galveston Psychoanalytic Institute, has supported and encouraged my work. My deceased mentor, colleague, and friend Dave Freedman, M.D., always helped me with key questions and observations. I miss David. My colleagues in my class at the Houston Galveston Psychoanalytic Institute (Jules Bohnn, M.D., Frank Gittess, M.D., Duane Purcell, M.D., Manuel Ramirez, M.D., Frank Aviles, M.D.) listened and gave feedback to me during the early stages of my thinking about the psychology of cults. Professor Jean-Francois Mayer was of enormous help in leading me to deeper understanding of the psychobiographies of Luc Jouret, Joe DiMambro, and the Solar Temple. Our collegial e-mail exchanges were invaluable. Dr. Alan Sapp was helpful in tracking down key information.

I am grateful for the excellent editorial help I received from Clair Aukofer, Dan Barlow, Dr. Gale Arrowood, and my son Nathaniel Olsson on early versions of this book. Susan Peery has been of inestimable help in the crucial and final phases of the manuscript preparation. Susan also provided ideas and support when I was stuck from various forms of writer's block and fatigue.

TABLE OF CONTENTS

INTRODUCTION

The Legend of the Pied Piper of Hamelin

Since medieval days in Germany, the legend of the Pied Piper of Hamelin has been passed down in folklore and tales. As the story goes, this memorable figure, popularized during the 19th century by poet Robert Browning, was hired by the town fathers of Hamelin to rid the town of a plague of rats and mice. Called "Pied" because of his two-colored costume, he skillfully played unusual and curiously irresistible tunes on his flute. The rats and rice, mesmerized, followed the Pied Piper to a river, and, still under his spell, went straight into the water to drown.

When the ungrateful citizens and leaders of Hamelin refused to pay the Piper his price as agreed upon, he was wounded and became enraged. In revenge, the Piper turned his focus upon the children of the village. He charmed them with his music and led them away into a magical mountainside that swallowed them, and they disappeared forever. The community of Hamelin was left to face profound grief, loss, and regret.

The term Pied Piper has become a descriptor for any person who entices and deceives others to follow him or her, especially to their doom. And as we will show, it is not only the childlike who are seduced by the music of this malevolent type of leader.

For more than twenty years, I have studied destructive and apocalyptic cult leaders like Jim Jones, David Koresh, Shoko Asahara, Marshall Applewhite, Charles Manson, and Luc Jouret and Joseph DiMambro. These cult leaders, the mesmerizing Malignant Pied Pipers of our time, led idealistic, father-hungry, or disillusioned young people away from their homes and toward destruction. Having an understanding of cult mentality and the pathological personalities of cult leaders is essential, for there are striking similarities between these deadly leaders and the newest example, Osama bin Laden and his Al Qaeda cult of ultimate terror.

The death toll from Jonestown, the Branch Davidian disaster at Waco, and other cults of the last 30 years is horrendous:

ESTIMATE OF DEATH TOLL

Jonestown (Jim Jones)	918
Tokyo (Asahara)	12
DiMambro/Jouret (Solar Temple)	69
Waco (Koresh)	90
Applewhite (Heavens Gate)	39
Manson & Family	40
TOTAL	**1168**

When you add in the thousands killed in bin Laden's terrorist attacks of September 11, 2001 (and those directed by him before and after that event), it shows the terrible power of apocalyptic cults and their leaders.

Although the study of cults is a natural fit with my professional training and experience, my personal motivation has always been more primary. When I was a boy and young man, my family attended a conservative evangelical Protestant church where "hell-fire and brimstone" preaching was no rare fare. It usually came in the form of charismatic visiting evangelists who would besiege our congregation for a long weekend. The majority of my experience in the church was of warm social acceptance, participation in church sports teams, and weekly worship services. In contrast, the episodic charismatic "Elmer Gantrys" were a source of intense fear, excitement, secret fascination, and disappointment for me. I was disappointed in my parents for exposing me and our family to these cathartic verbal humiliations and assaults.

Later, during my personal psychoanalysis, I came to realize that these intimidating strangers of the pulpit were also ambivalent heroes of mine. They commanded fear and wielded a great deal of power. It shook me to the core to realize that if Billy Graham had asked me to give up my life for Jesus, in my adolescent idealism I would very likely have swallowed the arsenic punch with fervor.

In essence, my work on this book takes some of its unconscious energy from my personal working out of my own early ambivalent encounters with preachers who wielded charismatic power over me in a way that would have paralleled the power of a Jim Jones or a David Koresh. I am deeply grateful to my analyst, who helped me to see this for myself. As a result, I have been able to help myself and my patients claim our freedom and autonomy to explore our own spiritual directions and not have them imposed by some guru (charismatic or otherwise).

Another personal impetus came out of a college friendship. When I was a freshman at Wheaton College in Illinois, I was on a soccer team with a sophomore named Tommy Stoen. Tommy's older brother, Tim, a senior, was class president and a brilliant pre-law student headed for Stanford Law School. Tim had a great sense of humor and I admired him. Much later, after the disaster at Jonestown, I was totally jolted when I read that Tim's son, little John Victor Stoen, had died at the side of Jim Jones. I later read the affidavit wherein Tim Stoen in essence gave legal permission to Jim Jones to father a child with his wife, Grace! As I read this riveting material in 1978, it began my years of research and writing about cults. *How could a brilliant, well-adjusted man like Tim Stoen fall under the spell of an Elmer Gantry-style jerk like Jones?*

In October of 1979, I won the Judith Baskin Offer Prize for a paper, "Adolescent Involvement with the Supernatural and Cults," published in *The Annual of Psychoanalysis* (Volume VIII/1980). As the first psychoanalytic work about cults (and in some analysts' minds, a "classic"), it got me invitations to speak at many professional meetings and led some people to consult with me when their children were drawn in by a cult. My life has been threatened on several occasions, especially when I included Scientology on my list of cults.

Throughout my professional life — more than 35 years — I have studied individual personalities as an empathic participant/observer at psychotherapy sessions and psychoanalytic hours. Many of these patients have been strong, charismatic, influential, and persuasive. Others have been dependent, longing and searching for father figures or parental approval, and (some observers would say) quite weak in character.

In addition to studying individual personalities and how to help them mature and change, I have also spent many years leading therapy groups and observing the powerful processes that arise in the life of the group. The psychological strength of any group working together is always stronger than the sum total of its members. This seems true whether the goal is a positive group accomplishment or a frightening group event like group-suicide, homicide (the Manson Family or the Tokyo gas attacks of Asahara), or war. A hard-working therapy group can facilitate cohesion, mutual support, and collective transcendence of a problem that often is impossible for an individual to solve alone. However, as we will discuss in this book, a group searching for easy answers to the meaning of life — an addictive affiliation with stimulation, excitement, and grandiose action — can be one of the most terrifying phenomena of the human condition.

Narcissism of Cult Leaders and Members

To understand the powerful pull of a cult, we turn to the psychological concept of narcissism. The dynamics of narcissism are poignantly applicable to our study of powerful cult leaders as well as to the loyalty and devotion found in their followers.

Regardless of our relative strength or weakness, most of us search for meaningful affiliation and self-esteem to build significance or meaning in our lives. Most of us gradually gain enough independence, wisdom, and ability to think critically so that our self-love is solid; we do not allow other people, no matter how charismatic, to dominate our core values or decision-making process. Some less individuated people do not seem to be so resilient at some phases of their life, for various reasons of circumstance, disposition, or chance. The psychoanalytic term for this complex area of self-love and dignity is narcissism, or the narcissistic sector of our personality.

It is important to note that narcissism is *not* a psychoanalytic curse word and is not a synonym for selfishness or self-absorption. In a normal personality, self-love matures and develops in a healthy way, just as one's ability to love others expands and matures.

It was Sigmund Freud who first observed that we love *anaclitically* (relating to the mother who nurtured us or the father who protected us) or *narcissistically* (relating to the self we wish we were, the self we used to be, or in affiliation with another self that reflects favorably upon us). Anaclitic literally means "leaning on," and refers to an infant's utter dependence on its mother or mother substitute for its sense of well-being and actual survival. Anaclitic love is normal behavior in childhood, but not in adulthood.

Narcissistic love is neither "good" nor "bad" in itself; an appropriate amount of narcissism is necessary for healthy self-esteem, empathy, and creative expression. Too much or too little narcissism interferes with a person's relationships with others: a deficit of narcissistic love often causes low self-esteem and feelings of shame or rage; an excess is associated with arrogance, entitlement, and self-centeredness.

When we mature and individuate (i.e., develop our own individual personalities, separate and distinct from all others) with relative freedom, we integrate the love we have found in our parental figures and move along in our own struggles and mistakes in loving. In this book, we examine how cult leaders and followers get seriously stuck in these core issues. And in destructive apocalyptic cults, we show how unloving the ultimate destiny of fanatical love can be.

The Cult Leader

The leader is crucial to any group. When a healthy group has an ethical, rational, and caring leader, it functions smoothly. A good or healthy spiritual leader helps facilitate rational decision-making within the group; exhibits incorruptible honesty; makes decisions with empathy and realism; shares leadership and cultivates it in younger members of the group; encourages individual freedom and dignity; respects members' families; and helps the group with positive projects that help the community at large. In times of great stress, healthy leadership matters even more. Think of New York City on 9/11 when leaders and people pulled together to help victims of the terrorist attacks.

In groups that lack an effective, healthy leader, external stress causes negative and regressive behavior. W. Bion, who studied group dynamics in the 1950s, offers valuable perspectives. He observed that in small groups (fewer than 15 members) under stress and without clear, effective leadership, members tend to fall into three main emotional patterns: dependency, fight-flight, and pairing. That is, they idealize the leader as omniscient and omnipotent and see themselves as immature and inadequate; they often are drawn to a paranoid leader who will lead the fight against external enemies; and they often focus on a couple who will (in the group's unconscious fantasy) survive the stress and ensure the longevity of the group. A cohesive group that feeds a leader's malignant narcissism and is fed hate in return only supersedes the self-absorbed, cruel, tyrannical, and sadistic qualities of the leader.

In larger groups (including whole communities and even societies) under stress and without a healthy leader, the relationship between leader and followers is not that of a benevolent shepherd and flock, but rather a reverberating, symbiotic "two-way street" in which members identify closely with each other, idealize the leader, and rally around him. In Waco or Jonestown, very few followers fled from Koresh or Jones. For another example of this "circle-the-wagons" reaction, think back to 1997 in Iraq. When the United States and its allies threatened action against Saddam Hussein, his followers formed a human shield around his palaces.

The Dangerous Lure of Cults

I share my personal and professional background because some colleagues and other professionals have occasionally accused me of being anti-cult. Many clergy and even experienced mental health professionals underestimate the scope and power of cult leaders and their lieutenants in controlling the minds of their recruits. For example, David Koresh, as we will see, did not utilize some innocent ideological salesmanship, but rather a powerful, repetitive, mind-controlling indoctrination over days and weeks and months that wore down a victim's psychic reserves to the point of snapping or *thought reform.* In bin Laden's Al Qaeda camps, in addition to military training, recruits received daily study of the Koran (something that attracted John Walker Lindh and others) and radical Islamist theology and indoctrination. The recruits were discouraged from communicating with their families. Recruits treasured their meetings with bin Laden, and many of them were given new names by the terror organization.

In 1983, Dr. John G. Clark, Jr. grasped the significance of this issue. In a paper entitled "On the Further Study of Destructive Cultism," he wrote that too many clinicians "tend still to explain cult conversions, as well as the difficulties arising from them, as results of long-standing personality or familial problems, as expressions of normal developmental crises, or even as manifestations of formal mental disorders. These observers tend to ignore the necessary role played by the cult milieu in causing the radical personality changes and family schisms that have clearly affected so many previously normal people and well-integrated families."

In a 1987 paper, I added a corollary: "This blurring of clinical and phenomenological cause and effect has led to much confusion and dangerous casualness about destructive cults as if they were the benign, relatively harmless variety." Cults provide a milieu where black-and-white thinking can grow and devour minds.

My study of cults has taken me on an interesting journey. For several years I had such counter-transference rage at Jim Jones that I had to stop writing and giving talks on the subject. My own analysis was very helpful in getting past this painful time.

It is my hope that this in-depth psychological study of destructive cult leaders of the last 30 years – Malignant Pied Pipers — illuminates the roots of their malevolence and their power, a condition that has invariably led to murder, mass suicide, the destruction of families, and to the terrorist acts that dominate our headlines. By understanding them and their appeal, we increase our chance of averting future disasters.

> *"Those who do not remember the past are condemned to repeat it."*
> — philosopher George Santayana, whose statement was glibly used by Jim Jones as a slogan above the "throne" he died on at Jonestown

CHAPTER ONE

UNDERSTANDING THE MALIGNANT PIED PIPER

"I was ready to kill by the end of the third grade. I mean, I was so fucking aggressive and hostile. I was ready to kill. Nobody gave me any love, any understanding. In those Indiana days a parent was supposed to go with a child to school functions ... There was some kind of school performance, and everyone's fucking parent was there but mine. I'm standing there alone. Alone, I was always alone!"
— Rev. Jim Jones, preaching in Jonestown, Guyana, one year before the mass suicides

"Dr. Olsson — They left me behind!"
—79-year-old Jonestown cult member and survivor of the apocalypse who had fallen asleep in her tent and awakened to find all of her fellow cult members dead

The Psychological Birth of the Malignant Pied Piper
How do charismatic / messianic personalities form? How does a charismatic / messianic person like Jim Jones or Osama bin Laden rise up from relative obscurity to the international media stage as a cult leader? What enables these Malignant Pied Pipers to play the role of idealized, worshipped, parental figures for their followers? What is involved in the psychology and group dynamics of cult leaders and their followers?

Psychologically, during their childhood or adolescence, these destructive cult leaders all experienced painful, life-altering disappointments in their parents. This occurred through neglect, abandonment, shame, severe disappointment, or humiliation. Each of the Malignant Pied Pipers I have studied has two key aspects to his or her aberrant personality development:

1. Childhood experience of protracted and overwhelming narcissistic woundedness. David Koresh, for example, was neglected and abandoned by his unwed teenage mother and father; Charles Manson's unwed mother rejected him, and she spent much of his childhood in prison. Jones also was a lonely and neglected child.

2. Later adolescent or young adult "dark epiphany" experiences echoing and compounding their earlier woundedness. These experiences — often rejection or abandonment — generally intensify an already malformed and rigid personality.

Not every person who experiences a traumatic childhood goes on to become a Malignant Pied Piper, but every Malignant Pied Piper I have studied started life with a love-starved, emotionally deprived childhood. For a sensitive and vulnerable child with a higher than average emotional IQ, which most of the cult leaders seem to have, these childhood wounds become insurmountable.

Diagnosis: Narcissistic Personality Disorder
It is impossible to completely determine the causes for the trajectory of any person's life. But every destructive cult leader I have studied, including Osama bin Laden, fits at least eight of the nine criteria for Narcissistic Personality Disorder. (See below.)

Osama bin Laden and the other Malignant Pied Pipers profiled here fit all four of the criteria for Malignant Narcissism (see below) described by psychiatrist and psychoanalyst Otto Kernberg in his book on severe personality disorders.

Definitions: Narcissistic Personality Disorder
Diagnostic and Statistical Manual (DSM-IV) of the American Psychiatric Association states the following:

> 301.81 Narcissistic Personality Disorder. A pervasive pattern of grandiosity (in fantasy or behavior), need for admiration, and lack of empathy, beginning in early childhood and present in a variety of contexts, as indicated by five or more of the following:
>
> 1. Has a grandiose sense of importance (e.g., exaggerates achievements and talents, expects to be recognized as superior).
>
> 2. Is preoccupied with fantasies of unlimited success, power, brilliance, beauty, or ideal love.
>
> 3. Believes that he or she is "special" and unique and can only be understood by, or associate with, other special or high-status people.
>
> 4. Requires excessive admiration.
>
> 5. Has a sense of entitlement, i.e., unreasonable expectations of especially favorable treatment or automatic compliance with his or her expectations.
>
> 6. Is interpersonally exploitive, i.e., takes advantage of others to achieve his or her own ends.
>
> 7. Lacks empathy; is unwilling to recognize or identify with the feelings or needs of others.

8. Is often envious of others or believes others are envious of him.

9. Shows arrogant, haughty behaviors or attitudes.

(Diagnostic and Statistical Manual of Mental Disorders, Edition IV, of the American Psychiatric Association, Washington, D.C., 1994. These criteria are virtually identical in the (2000) DSM IV-TR Quick Reference. Used with permission.)

Definitions: Malignant Narcissism
When a person already suffering from Narcissistic Personality Disorder becomes even more extreme in his or her beliefs and behavior, the condition is diagnosed as Malignant Narcissism. These people are the Washington, D.C., snipers, the Hannibal Lecters, Jeffrey Dahmers, Ted Bundys of our time; they also are the Malignant Pied Pipers that lead apocalyptic cults.

As described by Otto Kernberg, this condition involves four key elements:

1. Paranoid regressive tendencies with "paranoid micro psychotic episodes." These brief episodes of narcissistic rage involve loss of contact with reality and serve the function of punishing external enemies in order to avoid internal pain.

2. Chronic self-destructiveness or suicidal behavior as a triumph over authority figures. The malignant narcissist makes empathic followers or family feel his own hurt by initially seducing but eventually hurting them. Malignant Pied Pipers do this to their followers and even to their own children.

3. Major and minor dishonesty (psychopathy). Malignant narcissists manipulate and exploit others for profit, for their own satisfaction, or for imagined glory.

4. Malignant grandiosity with overt sadistic efforts to triumph over all authority. This triumph represents a satisfying turning of the tables for a malignant narcissist who, for instance, may have been abandoned without remorse by his father. By killing or vanquishing authority, the malignant narcissist feels as though he has achieved revenge against his uncaring father. (Kernberg, Otto, *Severe Personality Disorders: Psycho-Therapeutic Strategies,* 1986, page 195.)

Otto Kernberg's delineation of the malignant narcissist tunes in to the destructive cult leader — the Malignant Pied Piper — in an uncanny way. He says that the malignant narcissist's grandiosity and self-pedestalization are reinforced by the sense of triumph over fear and pain through inflicting pain on others, which brings an almost sexual pleasure in this interpersonal process. These types of narcissistic personalities exhibit a "joyful cruelty" (this was particularly true for Charles Manson) that allows them to obtain a sense of superiority and triumph over life and death through their own suffering and that of their chosen followers and victims. Destructive cult leaders invariably are heavily laden with malignant narcissism.

> *"You are stupid piss-ants and reptiles, who are lower than the primates... You fuckers, I like to look at you now, because you don't know how clever I am."*
> — Rev. Jim Jones, in one of his all-night sermons to his cult in Jonestown

Diagnosis: The Pied Piper of Hamelin
What do we make of the story of the Pied Piper of Hamelin from a psychoanalytic viewpoint? Any person with a normal amount of narcissism might become angry at the citizens of Hamelin for refusing to pay the agreed-upon price. But someone like the Pied Piper, a charismatic, grandiose figure, reacted much more strongly to the disappointment. He became enraged and in retaliation led the

children to destruction. A psychiatrist or psychoanalyst would diagnose this as Narcissistic Personality Disorder, or at least a reaction of pathological narcissistic rage.

The "Science," Philosophy, and Politics of a Psychiatric Diagnosis

During my study of destructive/apocalyptic cult leaders and their followers, I have struggled with the issue of applying psychiatric diagnoses (DSM-IV) to these individuals. None of them ever voluntarily consulted with me or any other psychiatrist. However, Charles Manson saw psychiatrists as a part of his incarcerations and court-ordered psychiatric evaluations during his murder trial. I met Marshall Applewhite on several occasions when I lived in Houston, but they were superficial social contacts. Applewhite had a psychiatric hospitalization and provided a note from a psychiatrist when he dropped out of an opera role in New York City. I assume he may have voluntarily sought help from a psychiatrist, but voluntary seeking of help is not a guarantee of meaningful therapeutic contact. Through the retrospection scope, I am convinced that Applewhite never completed any effective psychotherapeutic exploration of his personality or his inner conflicts.

In my experience, it is a rare and fascinating patient who knocks on my door because of external pressure from family members or a court order, and subsequently begins to look inward for insightful personal change and psychological understanding.

The Clinical Art and Science of Diagnosis

In my usual work as a psychiatrist, I see a patient for an initial evaluation interview and, if agreed upon, the patient and I schedule several more evaluation sessions. These three to six or more sessions allow both the patient to evaluate me and me to avoid premature closure or conclusions about the diagnosis (or diagnoses) or treatment plan recommendations. The evaluation phase also allows me to obtain medical and previous psychiatric or psychotherapy records with the patient's signed consent. If the patient under evaluation refuses this record review or my phone contact with previous treaters, this is valuable information in itself, and an important focus for our discussion.

For example, I recall one patient who was terrified at first to sign a release for records from a previous therapist who had sexually abused her. Our collaboration about this unethical psychiatrist allowed effective professional and legal confrontation with this loathsome "professional." On another occasion, the records from previous psychiatrist(s) allowed me to spot an unconscious pattern in the patient's difficulties with people that eventually allowed us to help him *not* to fire me just at the moment he most needed me to help him resolve the issue. I had pointed out that at just the moment that his former psychiatrists had confronted him with the key issue, he had grown uncomfortable and had fired them.

During the first several evaluation interviews with a new patient, I try to keep an open mind as a physician, psychiatrist, and psychoanalyst. I listen carefully for symptom patterns that may give indication of a medical condition that can mimic a psychiatric condition. I allow a freely hovering attention to what my patient says and what he appears not to be saying, because of fear, hate, erotic desire, or other powerfully repressed or suppressed emotions. I do not leap to quick conclusions; rather, I allow the memories, experiences, and reading about diagnosis over my thirty-plus years of experience to come to bear on the person I am evaluating. I make a sincere effort not to be judgmental, but I do not shy away from drawing professional conclusions from the data I am entrusted with by my patient. This process is frequently challenging, stimulating, and humbling. I have always regarded my work as a privilege and a sacred professional duty.

Often it is helpful for me to talk to a patient's internist or primary care physician to make sure that medical diagnoses that can mimic psychiatric symptoms can be excluded before we begin a course of psychiatric treatment. If I am uncertain about a diagnosis and therefore a treatment plan, I am not hesitant to recommend a second opinion with a respected colleague. Occasionally it is helpful for me to request psychological or neuropsychological testing to answer questions I cannot resolve at my clinical interviews with my patient.

Managed care and its accompanying superficialities have put tremendous pressure on psychiatrists to reduce the appropriate number of evaluation sessions for patients, because the core pressure is for a glib and brief evaluation and prescription of medication after one or two sessions. This pressure is mercenary, and it trivializes psychiatric care. I have tried to fight this pressure with varying degrees of success.

DSMs, Clinical Science, and Psychiatric Politics

The American Psychiatric Association's Committee on Nomenclature and Statistics developed DSM-I *(Diagnostic and Statistical Manual — Mental Disorders)* in 1952. Periodic revisions of DSM occur in collaboration with new versions of ICD (International Classification of Diseases), of WHO (World Health Organization). Beginning with DSM-III (1974), the DSM committee developed explicit lists of criteria and tried to stay neutral with respect to theories of etiology.

In the last "Decade of the Brain," psychiatrists have had less training and experience with psychoanalysis or psychodynamic psychotherapy. Neurochemistry, neurophysiology, and psychopharmacology have provided vast volumes of fascinating and important new information to be learned by a psychiatrist in training.

But many psychoanalyst psychiatrists feel that psychotherapy and psychodynamic theories have been neglected to the extent that psychiatry has lost its appreciation of mind in favor of brain chemistry. DSM-IV in particular was intended (and distorted) to be objective and able to be used for scientific and largely statistical studies. Many of us psychodynamic psychiatrists support the need for objective criteria for our science but also feel that in psychiatry, so many key factors involved with human behavior and the art of treatment can't be seen, measured, or quantified. Paradoxically, valuable diagnostic labels like "Ego Dystonic Homosexuality" or "Neurosis" have gradually been dropped from DSM for reasons of political correctness rather than through careful research.

However, in this book I chose to apply DSM-IV criteria, controversies and all, to destructive cult leaders. I believe DSM criteria and Kernberg's criteria for Malignant Narcissism helps us focus and order our thinking about these individuals. Strictly speaking, this is not a clinical use of DSM, but rather the applied psychiatric use to order the data about Malignant Pied Pipers that I describe in my methodology discussion, below.

Diagnosing the Malignant Pied Piper

My conclusion on the diagnosis is that all the Malignant Pied Pipers in my study have predominant characteristics of a Narcissistic Personality Disorder and the additional elements of Malignant Narcissism. I do not quibble with colleagues who favor or add the diagnosis of Antisocial or Psychopathic Personality Disorder, because there is a lot of evidence to support that idea. Some observers can point to grandiose, paranoid, and delusional qualities in the thinking and teaching of all the Malignant Pied Pipers I discuss.

I do not deny these phenomena, but the pattern of Severe Personality Disorder ("crazy like a fox"), rather than psychosis, seems to me to be the more enduring diagnostic issue. People with severe personality disorder (as opposed to anxiety disorder or neurosis, depression, and so on) rarely come to a psychiatrist with the experience of inner pain from guilt or shame. They always externalize or blame other people or circumstances outside themselves for their difficulties. This is certainly true of Malignant Pied Pipers.

I also would not deny that at times substance abuse was a factor, particularly with Jim Jones in the later phases of his personality disorganization. Jones's drug use preceded and then merged with paranoid and depressive elements as the apocalypse neared.

Other Useful Definitions

Personality: "The characteristic way in which a person thinks, feels, and behaves; the ingrained pattern of behavior that each person evolves, both consciously and unconsciously, as his or her style of life or way of being." (*American Psychiatric Glossary,* 7th edition, 1994, page 98.)

Personality Disorder: "Enduring patterns of perceiving, relating to, and thinking about the environment and oneself that begin by early adulthood and are exhibited in a wide range of important social and personal contexts. These patterns are inflexible and maladaptive, causing either significant functional impairment or subjective distress." (*American Psychiatric Glossary,* page 99.)

Antisocial Personality Disorder: "In the older literature called *psychopathic personality;* descriptions tend to emphasize either antisocial behavior or interpersonal and affectional inadequacies, each at the expense of the other. Among the more commonly cited descriptions are superficiality; lack of empathy and remorse, with callous unconcern for the feelings of others; disregard for social norms; poor behavioral controls, with irritability, impulsivity, and low frustration tolerance; and inability to feel guilt or learn from experience or punishment. Often there is evidence of conduct or disruptive behavior in childhood or of overtly irresponsible and antisocial behavior in adulthood, such as inability to sustain consistent work behavior, conflicts with the law, repeated failure to meet financial obligations, and repeated lying and 'conning' of others." (*American Psychiatric Glossary,* page 99.)

Methodology of This Book
I have not personally interviewed or examined any of the destructive cult leaders described in this book. (As I mentioned earlier, I had several casual social contacts with Marshall Applewhite in Houston many years ago.) Given a Malignant Pied Piper's bent of personality, it is unlikely he or she would ever voluntarily seek psychiatric or psychotherapeutic help. In fact, one of my reasons for writing this book is that these destructive cult leaders have such authoritarian and non-introspective personalities that they routinely deceive and con unsuspecting victims. They need to be spotted for who and what they are, because they exude supreme confidence and certainty on the surface of their smooth charm.

Some psychoanalysts such as Kohut and Kernberg have worked with highly charismatic-messianic individuals and have provided compelling observations and theories to shed light on these kinds of personalities.

My diagnostic conclusions and speculations are founded on a depth and density of thoroughly studied secondary sources. Many exploitative cult leaders love to talk, write, be videotaped, and be interviewed. During their sermons, pontifications, "healing" services, and "inspired" writings, they become quite emotional, autobiographical, and self-disclosing about their attitudes, passions, and motivations.

I have listened to many hours of Jim Jones's extemporaneous sermons in the Guyana jungle, where vivid self-disclosures revealed Jones's inner demons, conflicts, and fears. In fact, I experienced similar responses to hearing Jones's "White-Night" informal sermons and rantings as I did listening to the emotional self-disclosures of my psychoanalytic patients. My inner anger or sadness (counter-transference feelings) has been so intense at times that I have had writer's block for months at a stretch. (Counter-transference feelings are the complicated personal emotional reactions that therapists have toward their patients as the patient discloses painful, cruel, or vivid emotions.)

These reactions on the part of the therapist usually stem from unresolved problems of the therapist or have been resolved by the therapist during his or her own personal analysis. The reactions indicate profoundly painful, yet important areas of the patient's conflicts that would never be explored or resolved if it were not for the therapist's refusal to be frightened off from the exploitation of such heuristic but challenging therapeutic work. At times I have felt such profound loathing for Jim Jones that I wished he would be resurrected, so that I could beat him to a pulp! At other times, I have wished that he had been my patient as a young man, when I might have worked with him towards resolving his loneliness and hate so that he never would have continued to travel down the road to apocalypse as a terrifying pseudo-resolution.

Writers, journalists, and law enforcement or political figures have written about, investigated, and described the life events and life histories of cult leaders. While many critics cast doubt on the veracity of these sources, I would point out that the depth and density of the data produced by journalists such as Tim Reiterman in *Raven* or James Reston Jr. in *Our Father Who Art in Hell* is invaluable because of their extensive interviews with people who knew Jones in or outside of the spotlight of adoration or loathing.

The observations of devoted followers or the wounded perspectives of ex-cult members both have obvious sources of bias, but nevertheless have some value in describing the allure of a cult leader's program or its distasteful qualities.

The families and friends of the cult leader or the cult followers may be even more biased, but they offer valuable data on love and hate that brings all of us closer to the emotions of this subject. The reader will notice that the greatest volume, depth, and density of psychological data in my work is focused on Jim Jones. The reason for this is the great depth of material available to my searches about Jones as well as my acquaintance with Tim and Tommy Stoen years before Tim's terrible seduction at the hand of Jones.

Unconscious motives of individuals and groups are particularly difficult to demonstrate. But in these group-death events, the unconscious motives of the leader and the collective unconscious of the cult group is of profound importance in allowing us to make significant and persuasive inferences about Malignant Pied Pipers.

What Is a Cult?

Religious social movements are generally seen as a consequence of the failure of main-line religions to meet adequately the spiritual, existential, or other transcendent needs of a segment of a large population of people within a society. But not all social or religious movements give birth to cults; and not all cults spring directly from those movements.

In distinguishing a cult from other social or religious movements, we need to identify the elements that give a cult its special intensity or identity.

* The *American Psychiatric Glossary* chose conciseness in defining "cult" as "a system of beliefs and rituals based on dogma or religious teachings that are usually contrary to the ones established within or accepted by the community."

* The *New Oxford American Dictionary* adds the sense of "a misplaced or excessive admiration for a particular person or thing."

* Alexander Deutsch emphasizes the leadership element in his effort towards a definition: "A cult can be defined as a relatively stable, transcendentally oriented group surrounding a powerful leader who influences his or her followers in a direction that deviates strongly from that of the dominant culture." ("Tenacity of Attachment to a Cult Leader: A Psychiatric Perspective," *American Journal of Psychiatry,* 1980, Vol. 137, #12, pages 1569-1573.)

In her introduction to *Cults in Our Midst* (1995), Margaret Singer astutely observes that many writers and scholars avoid the term "cult" because of implications of weirdness or abnormality. The word "cult" is commonly used in a pejorative, not merely descriptive, way. In fact, as Singer has found in her interviews with more than 3,000 current or former cult members, they are far from marginal and the majority of those who join are the same as all of us. Singer states: "The issues they represent are basic to our society, to our understanding of each other, and to our accepting our vulnerabilities and the potential for abuse within our world." (Singer, Margaret, *Cults in Our Midst,* 1995, page xix.)

Cults — Bad or Benign?
Singer develops a broad-based classification of cults. She says they are many in number, and range on a continuum from benign groups to those that exercise extraordinary control over their members' lives. What is unique about Singer's approach is that she locates a

dynamic triad of "cultic relationships," and not just a static "definition."

The three dynamic factors are:

1. The origin of the group and role of the leader.

2. The power structure, or relationship, between the leader(s) and the followers.

3. The use of a coordinated program of persuasion, called thought reform or brainwashing. *(Cults in Our Midst,* page 7.)

Singer describes 10 major types of U.S. cults: 1. Neo-Christian religious; 2. Hindu and Eastern religious; 3. occult, witchcraft, and Satanist,;4. Spiritualist; 5. Zen and other Sino-Japanese philosophical mystical; 6. racial; 7. flying saucer and outer space phenomena; 8. psychology or "psychotherapeutic"; 9. political; 10. self-help, self-improvement, and life-style systems. *(Cults in Our Midst,* page 14.)

I add an additional category: the purely criminal or psychopathic cult. The Manson Family falls in line with the destructive cults we are studying. Charles Manson's "Helter Skelter" cult was characterized both by the homicidal acting-out of Manson's apocalyptic ideology and the leader-follower dynamics so typical of malignant cults.

Singer's work is valuable as a baseline and reference point as we explore destructive cults, because she helps us to avoid "demonizing" or "pathologizing" cult leaders or groups. However, she does not avoid the harsh clinical and social-psychological realities that cults present in our "free" society.

The Benign Cult
Rather than focusing immediately on the pathological domain, it is helpful to start with a description of the activities and shared experiences of a psychologically healthy religious group.

* A healthy religious group has centrally shared experiences of worship, reverence, wonder, and humility. In this context, the leader always shares leadership with various group members.

* The healthy group gives positive support for life-long loving connection, communication, and support for and from the family of origin and extended family.

* Respect and support for the sanctity and self-respecting boundaries for marriage and sexual commitments for couples is found in the group's respect for freedom and individuality in these choices.

* The healthy congregation embraces projects that contribute to the betterment of the community at large and not just the religious group itself.

* Finally, the money gathered by the group is used to pay the salaries of parish staff and buildings and is then contributed to projects to help the community or society.

There are some benign, often quirky or kooky, quasi-religious gatherings and sects that are essentially harmless. In addition, personality cults sometimes form around popular and charismatic celebrities and media stars, such as Dr. Phil, Oprah, John Bradshaw, and others.

The Bad Cult
Exploitive and destructive cults have many characteristics in common.

* They concentrate a large percentage of their group efforts on fund-raising, recruiting new members, and

subtly or obviously controlling the financial, social, familial, and sexual lives of their members.

* Cult members are frequently lectured at or berated by the charismatic leader after the initial "love bombardment" of recruitment or seduction. (Langone, page 7 and pages 98-99.)

* The destructive cult devalues or attacks the nuclear family or extended family. "The Family," a California cult, repeatedly said to recruits, "Spending time with your biological family is like eating your own vomit!"

* Destructive cult leaders often rearrange marriages, dictate sexual boundaries, or personally select sexual partners for themselves from the membership.

* The exploitive cult rapidly develops "in-group, out-group" dynamics, with group-sanctioned isolation from other groups and especially from a free range of dialogue about the differences in opinions between groups.

* Money gathered by the cult group is often used to enhance the fame, power, or prestige of the leader and not to primarily help the community.

* These cults have doctrines and practices based on a living leader's evolving revelations, which supplant or devalue rather than supplement or deepen traditional teachings.

Thus these groups retain as their central goal the aggrandizement and narcissistic affirmation of their charismatic leaders. They actively promote isolation from the rich variety of ideas abounding in the broader society. Their leaders seek to alienate members from their families or communities of origin. When two patients of mine, the parents of two daughters ensnared by Ramtha, tried to get high school friends, teachers, or hometown clergy to write to their daughters, the letters were returned unopened. In fact, the cult group is often literally as well as figuratively inserted as a new member's "new and superior family."

Case Study: J. Z. Knight, the Pied Piper of Ramtha

In 1987, a well-dressed couple in their mid sixties arrived at my office for consultation. John Abrams's ready smile and calm, confident demeanor was evidence of his experience with many kinds of people in the business and social worlds. Helen Abrams was frowning, worried, and sad. Their family physician had recommended me to them on the advice of his parish priest, who had heard me talk about destructive cults.

Mr. Abrams began. "Doctor, my wife and I are beside ourselves. Our daughters have both had their lives and souls taken over by J. Z. Knight and her Ramtha cult in Washington. Both of our daughters had trust funds of one million dollars each. The trust monies became theirs when they were twenty-five years old. Beth is now 35 and Sara is 33. They were welcomed by Ramtha with open arms, and so was all their money. Our trust attorney notified us that large sums were going to Ramtha, but our daughters could not be dissuaded. They are devoted to Mrs. Knight and her Ramtha cause.

"Dr. Olsson, both of our daughters are bright, talented, and savvy. When Beth's husband died of cancer three years ago and Sara's husband asked her for a divorce about the same time, that's when it all started."

Helen Abrams burst into the conversation. "Ramtha has a lovely looking childcare center and our daughters were glad they found a place in Seattle where they felt their children were safe. Sara told me, 'Mother, the Child Haven is filled with love, and there are kind men there for my fatherless kids. Mother, I feel so comfortable with the Ramtha followers, and Mrs. Knight's talks are fascinating. She has a dramatic style. It is our home away from home.'"

Helen was crying now. "We have four grandchildren and those 'kind' Ramtha people won't allow Sara and Beth to let us see them! As soon as our daughters gave most of their money to Ramtha, they saw less and less of J.Z. Knight. Sara and Beth had given up their public-school teaching jobs and were expected to teach at Ramtha schools. Their tone of voice over the phone grew detached, less spontaneous, and John and I felt Ramtha people were present during the calls. We were never allowed to talk to our grandchildren."

Alarmed, the Abramses had flown to Seattle. They found armed guards at the gates and throughout the Ramtha compound. They were allowed to see their daughters only briefly and in the presence of Ramtha "counselors." When the Abramses protested, the counselors said the daughters had not yet reached the "level" of private visits. Both daughters were strangely detached, passive, uncritical of the counselors, and cool to their parents.

When John Abrams became angry at the situation, the guards ordered him and Helen to leave. Fearful of a confrontation, Helen practically dragged her husband out the door. She related, "I will never forget the looks on our daughters' faces when I looked back. They wore mixtures of blank, perplexed, half-smiling frowns. It was eerie, sad, and frightening at the same time – like an old zombie movie."

The Abramses stayed in the area for a week, having several brief, supervised visits with their daughters but never with their grandchildren, who were said to be in special classes or napping. They went to the local police, who said that the daughters were at Ramtha voluntarily and the police could not intervene. They even spoke with a deprogrammer, but after meeting with him decided that the deprogrammer's approach – essentially kidnapping all six of them — would be too traumatic for the grandchildren.

They also attended several public talks by J. Z. Knight. Each talk was preceded by rhythmic singing, chanting, and instrumental music. Then J. Z, Knight, a tall, attractive blonde, began speaking. "She began each talk with teachings about reptilian creatures she said were located on the dark, invisible side of the moon," John

Abrams reported. "These creatures were planning to take over the Earth. She said that the U.S. government refused to admit the gravity of the threat, so Ramtha leaders were preparing the defenses. Doctor, it all sounded so glib and absurd! But all of a sudden Knight's voice grew deep and she became Ramtha, saying, 'I am a fifteenth-century warrior! You are all in danger and must listen to what I say.'

"She went on and on about purifying their personalities and strengthening their character for the ordeals to come, like some evangelist talking about Armageddon in a hell-fire and brimstone revival meeting. Then she suddenly rubbed her eyes and became Mrs. Knight again."

Helen Abrams looked over at her daughters and saw that they both were spellbound by J.Z. Knight's transformation to Ramtha, a male warrior from 35,000 years ago. She sobbed to see her girls so mesmerized by a charlatan. Recently, the Abramses had seen newspaper articles about Ramtha that described how J.Z. Knight had expanded the cult aggressively when her husband sued her for divorce because of her sexual relationships with some of her male lieutenants. J.Z. Knight had claimed that her husband needed money because he had contracted AIDS. Helen and John tried to show their daughters the articles, but no magazines, newspapers, or other media were allowed in the compound. John Abrams concluded their description by saying, "Doctor Olsson, it is like our daughters and granddaughters are under the spell of a malevolent pied piper. We need your help!"

I worked with the Abramses for two years. John's anger about losing two million dollars was huge, but even more powerful was his shame and humiliation that his two educated daughters had been duped. I supported the parents' efforts to keep in contact with their daughters. They sent letters, clippings, pictures, and other communications frequently, and at my suggestion kept copies of everything they sent. Friends, relatives, former teachers, and mentors of the girls also wrote to them. No one received any reply.

A break came after 14 months. Beth's older daughter, Shelly, became very ill with a high fever and had a seizure. The Ramtha people insisted that Beth wait to see the Ramtha doctor, who wasn't available for several hours. In a panic, Beth managed to sneak out of the compound with both of her girls and hitchhiked to a local emergency room. She had no money, and telephoned her parents in the middle of the night. John immediately wired money and he and Helen were on the next plane.

Shelly had encephalitis, and during her long recovery John and Helen spent hours showing Beth the copies of the letters and pictures they had sent – and which she had never known existed. Ramtha counselors came to the hospital and tried to get Beth to return, but with her parents' support, she was able to muster the strength to refuse.

At my sessions with the Abramses, we had often discussed how delicate the separation from the power of Ramtha might be. They realized how important it was for them not to "kill the Ramtha messengers," who were, in fact, their daughters.

I saw Beth for many individual therapy sessions as she gradually allowed her strength of personality and sense of humor to return. She revived old friendships to, as she put it, "Perform my own caesarian section from Ramtha." It took more than a year for Beth to move beyond the uncritical passivity, naïve dependency, and blunted emotions that her cult experience had brought. On one occasion, a call from a Ramtha counselor came when she was home alone, and she almost returned, but calls to her father and two friends helped her to "overcome that magic voice over the phone."

In 1990, Beth and an old college friend of Sara's were able to reach Sara on Christmas Eve while her Ramtha counselors were at a meeting. Beth told Sara about her hospital experience, and she and the friend read letters and greetings from old friends of Sara's. At first Sara was defensive, but when she talked to Shelly, Sara began to cry. Shakily, she was able to leave the compound with her daughters and get a ride to the airport.

Sara's therapy was more difficult than Beth's. Sara had been made a junior counselor and head art teacher at the Ramtha school, and her daughter Brittany was a Ramtha school leader. A child psychiatrist was helpful to Sara's daughters, especially when they returned to public school and tried to convince their classmates about the lizards on the far side of the moon. Work with the girls and their teachers helped to address their peculiar combination of grandiosity and uncritical passivity about Ramtha.

Diagnosis: J.Z. Knight's Appeal

J. Z. Knight, a "channeler," has been particularly adept at recruiting middle-aged and older women who seem to feel vicariously empowered in her presence. Ramtha's apocalyptic scenario – a pagan fight against evil forces led by the ancient warrior Ramtha (Knight) — is not deadly, but financial and emotional theft and exploitation run rampant there. The cult is still active.

It could be argued that J. Z. Knight has Multiple Personality Disorder (DSM-IV calls it Dissociative Identity Disorder), common in persons with a history of early childhood abuse. But the main diagnosis in my opinion is Narcissistic Personality Disorder with hysterical or histrionic elements.

Beth and Sara's recruitment, when both were at vulnerable times in their lives, illustrates Margaret Singer's idea that all of us are susceptible at certain "in-between" times in our lives. The strength of the Abrams' women's commitment to Ramtha shows the dynamic between the searching in-betweener and the charismatic Pied Piper. Both Sara and Beth confided in me that it was of jolting importance to them to learn that their dad loved and cared about them in addition to his money!

Malignant Pied Pipers and Their Followers

Let me reiterate the core thesis of my observations about exploitive-destructive and apocalyptic cult leaders, whom I call Malignant Pied Pipers. They seem to develop a relentless quest to become strong parents or father/mother figures for themselves. This lifelong search, of course, requires a ready supply of child-admirers, who are found in the cult followers. Malignant Pied Pipers give psychological birth to followers through an array of indoctrination and seduction techniques.

Apocalyptic-destructive cult leaders have experienced harsh disappointments in their own parents and often their home community via neglect, abandonment, shame, or humiliation during their childhood. As the years go by, their loneliness and their memories of empathy-starved and shame-dominated childhoods become magnified, as if these lonely, humiliating years had become a psychological deformity.

These poignant experiences of neglect, shame, psychological abandonment, and fear of being alone lead them toward dramatic action. Actions provide a sense, however spurious, of inner restitution and parallel revenge. The evolving dynamic in the cult family group is a two-way street between the members' passive or masochistic narcissism and the active, aberrant behavior advocated and orchestrated by the leader. This time the leaders feel empowered, rather than powerless as they were in childhood. In essence, the cult leaders gain a sense of power and mastery over their own early childhood feelings of insignificance by becoming overwhelmingly significant and powerful in the lives of their cult followers.

Beneath the outward confidence and swagger of the leader is an unconscious sense of shame and a fear of humiliation ready to surface when the leader's progressively fragile narcissism is punctured. Thus the cult-family honeymoon is eventually over; and because of the leader's own malignant narcissism, he recapitulates the neglect, abuse, and ultimate loneliness and victimization he once experienced. But, this time, the cult leader is the neglecter, abandoner, and victimizer. And the grand finale is suicide or homicide.

The Dark Epiphany of the Malignant Pied Piper
Modern psychoanalysts have recognized that adult experiences are often as important as experiences in the first five years of life. The internalized experiences with mothers, fathers, siblings, grandparents, and extended family members are formative and important. Also important are experiences with peer friends and enemies, teachers, mentors, coaches, employers, coworkers, and paramours, as well as

national events, monuments, and cultural myths or rituals.

The future destructive cult leader has had significant young-adult molding or formative experiences that equal and usually go beyond the intensity of what has been called a "mid-life crisis." I call these experiences of future destructive cult leaders "dark epiphanies." Each biography in the chapters that follow will illustrate that leader's dark epiphany (or epiphanies) and show how the experience acted as a trigger or turning point, propelling the cult leader toward a more focused and desperate desire for power, control, and especially action.

For example, when Jim Jones was rejected by Father Divine's congregation (after Divine's death), he became obsessed with becoming a god himself and finding a safe place for his cult. The dark epiphanies confirm and magnify the worst childhood experiences and make the leader's apocalyptic climax even more urgent.

The Apocalyptic Death Vision

> "I don't mind losing my life. What about you? I don't mind losing my reputation. What about you? I don't mind being tortured. What about you? ... I'm no longer afraid. I've lost interest in this whole world of capitalist sin ... I'd just as soon bring it to a gallant, glorious screaming end, a screeching stop in one glorious moment of triumph."
> — Rev. Jim Jones, six years before the Jonestown mass suicides (Reiterman, Tim, with Jacobs, J., *Raven: The Untold Story of the Rev. Jim Jones and his People,* 1982, p.293.)

The most extreme forms of the scenarios of spurious restitution and revenge in cults, such as the Jonestown disaster, are the mass group suicides or homicides. These dramatic climaxes come out of the cult's shared vision of apocalypse, which is spawned and cultivated in the group's psychological life by the leader. The leader repeatedly, boldly holds forth the noble and treasured fantasy of dying together for "The Cause." The Malignant Pied Piper obtains inner feelings of superiority, revenge, and an empty restitution, which is the epitome of the denial of death itself. The apocalyptic vision also provides an ongoing drama that keeps emotions at a high pitch and gives the cult members feelings of power, importance, or destiny. Their lives are no longer humdrum or boring, yet they can remain passive participants under the leader.

In the case of the duped followers of bin Laden's illicit *jihad,* they are promised special favor with Allah. They become brainwashed and believe that they will be accorded a favored status of pleasure in Allah's presence in heaven immediately after their deaths as holy martyrs – even if their suicidal/homicidal act involves the death of innocents. Bin Laden grandiosely designates America and Israel as the evil entities to be destroyed, even at the cost of his followers' lives via suicide bombs.

If these leader-follower dynamics permeate the early and middle phases of the cult life, what happens within the leader to activate the apocalyptic group-death scenario? I propose that the leader experiences a midlife malignant epiphany or series of malignant epiphany experiences. The dark epiphanies are cumulative and on-going events. In Erik Erikson's epigenetic model (*Childhood & Society,* 1950, pages 266-268), these experiences create isolation, stagnation, and despair, rather than intimacy, generativity (nurturing of the next generation), and ego integrity, in their young adult, adult, and older years. The apocalyptic group death is not a sudden, impulsive command from the leader at a time of crisis. Group death has been woven into the fabric of the cult as a dramatic, defining myth. It becomes a source of paradoxical heroism and cohesion that magnifies the special domain of the leader and makes him needed even more.

The Malignant Pied Piper does not reach old age. Strong, dramatic group action provides the ultimate, fantasized triumph over the leader's own insecurity and phobic dread of helplessness and isolation. The leader acts out his self-hatred by leading the cult group into death. Recordings of Jim Jones's final rantings at Jonestown in Guyana, with his followers' comments in the background, reveal the poignant codependent dynamics of doomsday.

> *"I can do anything I want because I've sacrificed to give everybody the good life. I come with the black hair of a raven. I come as God Socialist."*
> — Jim Jones, in a ham radio broadcast from Jonestown

CHAPTER TWO

REV. JIM JONES AND THE PEOPLE'S TEMPLE OF JONESTOWN

"I do not believe in violence. Violence corrupts. And they say I want power. What kind of power do I have walking down the path talking to my little old seniors. I hate power. I hate money. The only thing I wish now is that I was never born. All I want is peace. I'm not worried about my image. If we could just stop it. But if we don't, I don't know what's going to happen to twelve hundred lives here."

— Rev. Jim Jones to Congressman Leo Ryan, in response to a question about brutality toward cult members, one day before the apocalypse at Jonestown, 1978

Biography

James Warren Jones was born on May 13, 1931, in Crete, Indiana. His father, James T. Jones, had been a road construction foreman and was known as a good man before serving in World War I. During the war, mustard gas scarred James's lungs, leaving him a sickly, broken man. He rarely worked and got by on public welfare. It was rumored he belonged to the Ku Klux Klan. He had very little interaction with his son, Jim. Lynetta Jones, Jim's mother, was a slender, energetic woman who smoked, drank beer, worked hard, and in effect raised her son alone, reserving her scorn for her husband. Lynetta Jones was skeptical of organized religion but believed in "spirits," a belief she instilled in her son.

In *Raven,* author Tim Reiterman (*Raven,* pages 16-17) describes a neighbor woman named Myrtle Kennedy who was kind to and protective of little Jim Jones. Mrs. Kennedy was an avid fundamentalist who was determined to save his soul. Jones would often roam the neighborhood with his friend, Don Foreman. Two or three times a week, six-year-old Jim would stop by a car garage and entertain the men who hung around there in order to get soda pop. (Lynetta Jones labeled those men "the loafers.") Jim would cut loose with a string of cuss words and called them dirty bastards and sons of bitches. The "loafers" loved it and bought him and Don sodas.

In his early school years, Jones fell in love with books and soon was reading several levels ahead of his grade. He was a B student, outspoken in class and sure of himself. Reiterman described Jones as a "roguish little natural leader." Reiterman says further, "Outside school, he could control the same playmates who intimidated him at school. He structured the environment to suit himself, using a certain knack that, when full-blown in adulthood, could rightly be called genius. He learned at a very early age how to attract playmates, keep them entertained and maintain a hold on them. To accomplish it, he shifted modes, from playmate and companion to dominator, pushing his authority and then backing off." (*Raven,* page 17). Young Jones loved dogs and even conducted funerals when pets died (*Raven,* page 19). As a teenager, he delved into socialist and communist literature.

In September of 1954, at age 23, Jim preached at a Pentecostal church in Indianapolis, advocating a gospel of racial tolerance and socialism. The next year, he started a new church, called Wings of Deliverance, soon renamed the People's Temple. During the 1950s Jim became fascinated with a charismatic black preacher named Father Divine, whose congregation and wealth were growing exponentially. Jones took busloads of his young followers to Philadelphia to hear Father Divine, and mimicked his frenzied style of preaching and raising money.

When Divine died in 1965, Jones was rejected as the new leader of Divine's church, an event that would figure as one of his "dark epiphanies." Stung, Jones proclaimed that Divine — along with Jesus and Lenin — had entered his soul. His People's Temple, loosely affiliated with the denomination Disciples of Christ, came under heavy criticism in Indiana for its radical theology, so Jones decided to move his church to Ukiah, California, taking some of Father Divine's adherents with him. Jones believed Ukiah was isolated enough to be a haven for racial equality and safe from the threat of nuclear war (a preoccupation of his). In California, Jones consolidated his power, recruiting affluent professionals with his gospel of social equality while funneling their money into his control.

In 1974, Jones leased land in Guyana for a colony and sent a few members of the People's Temple to start clearing the jungle and building Jonestown. In 1977, when the Internal Revenue Service began scrutinizing the People's Temple for Jones's practice of commandeering elderly parishioners' Social Security checks (to the tune of $65,000 a month by early 1978), Jones decided to move operations to Guyana. Many of his followers moved there.

During 1978, word began filtering back from defectors that Jones was brainwashing the faithful in Guyana, holding them against their will, abusing children, and becoming more and more deranged. Several high-profile defectors, including Timothy Stoen (once Jones's second-in-command), convinced California Congressman Leo Ryan to investigate.

Ryan, with a group of reporters and photographers and several concerned family members, flew to Jonestown in November of 1978. On the second day of the visit, as they attempted to leave Jonestown, they were ambushed on the airstrip by gunmen sent by Jones. Congressman Ryan, three members of the press, and one Temple member who was attempting to leave Guyana were killed.

Back at the Temple, Jones assembled all of his followers and proclaimed that the end had come and "revolutionary suicide" was the only option. A lethal mix of potassium cyanide, sedatives, and purple Fla-Vor-Aid was mixed in a vat. They started with the babies. Within a few hours, 913 people died of poisoning, including 276 children. They also poisoned their dogs. Jim Jones, the only one who didn't take poison, died from a single bullet to the right temple. His body was found near his throne.

> *"The more selfish a person, the more poignant his disappointments. It is the inordinately selfish, therefore, who are likely to be the most persuasive champions of selflessness. The fiercest fanatics are often selfish people who are forced to lose faith in their own selves."*
> — Eric Hoffer

How was Jim Jones able to lead more than 900 people into self-destruction in the Guyana jungle? What complex group dynamics permeated his apocalyptic cult? What is involved in the deadly symbiosis of cult leader and follower?

Jim Jones's Psychological Roots

Jones's emotional, self-revelatory sermons were like a rambling stream-of-consciousness catharsis before his audience of "true believers." One conclusion I draw from these sermons is that Jones, like David Koresh, had such a lonely, severely narcissistically injured, and empathy-starved childhood that he developed a deep phobia of dying alone. Reacting to the loneliness, neglect, and abandonment of his childhood as if it were a congenital psychological deformity, he developed extreme feelings of entitlement, acted out upon and with the cult.

Jones's father was a defective, distant, and sickly role model or father figure to his only child. He collected government checks (just as Jim Jones later would scoop up the pensions of his elderly cult members). He was an object of pity who rarely smiled. Too sick to work steadily, he often spent his days at a pool hall.

Jones's mother, Lynetta, was rarely home to take care of her son. She never went to church or to his school programs; she did not sleep with her husband or show him empathy or affection. She did encourage Jim to study hard, and harped at him, "Don't be nothing like your dad. You have to make something of your life and be somebody. Work at it. Nobody's gonna help you" (quoted in Reiterman and Jacobs). She also boasted of her son's future glories, comparing him to Albert Schweitzer and imagining her own reflected glory. In the end, she followed her son to Jonestown and died there before the Apocalypse.

In constituting his own family, Jim Jones seemed to bend over backwards to create an atmosphere of unconditional love and acceptance he had never felt as a child. Soon after he graduated from high school, Jones married Marceline, a nurse at Reid Hospital, where he was an orderly. She was four years older than Jim and had strict religious beliefs. Almost from the beginning she considered divorcing him because of his dominating and merciless behavior toward her, his dogmatism, and his obsessive need for power and control. Perhaps Marceline's love for her children and her religion bound her to her husband, but her increasing complicity and victimization became the pattern of everyone whose life became entwined with Jim Jones.

Jim and Marceline had eight children, a son of their own named Stephan and seven adopted children. Two of the adopted children were black and three were Korean. The Joneses called themselves the Rainbow Family and claimed to be the first white family in Indianapolis to adopt a black child. In a newspaper interview, Jones said, "Integration is a more personal thing for me now; it's a question of my son's future" (quoted by L. Wright in "Orphans of Jonestown," *The New Yorker,* Nov. 22, 1993).

In the 1950s, Jones's ministry and ideology was a mixture of liberal Christian social gospel and emotional faith healing. In the early days in Indianapolis, Jones ran soup kitchens for the poor and elderly. He always championed racial integration in the segregated city, where the KKK had marched in the 1920s and crosses were still being burned on lawns in 1960. Reiterman says, "Helping the disadvantaged was foremost in his mind, but under that was a personal need to be admired, loved, and lauded by the crowds."

In the 1960s, Jones became pervasively phobic about nuclear holocaust, and searched around the Western hemisphere for the safest location for his flock. He finally chose northern California. He appeared to be struggling externally with a vague awareness of his own internal explosiveness. In the early days of his ministry he sought out older clergymen to act as quasi "therapist" father figures. But as soon as they gave him feedback or criticism, he grew defensive and rejected them. Jones only revealed his insecurity, suspiciousness, and hypochondria to his wife, who used her nursing skills on him during his episodic "collapses" in their home when he was under stress.

Slowly and relentlessly over the 1960s and early 1970s, Jones sought powerful political connections by mobilizing People's Temple members to swell the numbers at political rallies. Glowing letters of gratitude were sent to him by Rosalyn Carter and Walter Mondale during the 1976 election campaign.

The chicanery involved in his faith healing was very clear, however, and reports of Jones's cruelty toward and sexual exploitation of his flock began to draw media attention. The move to Jonestown in Guyana was partially to stem the wave of defections by cult members, and also to avoid media investigations and IRS scrutiny of Jones's use of members' Social Security money. It also reflected Jones's paranoia and narcissistic personality deterioration.

In describing charismatic-messianic personalities like Jim Jones, Kohut observed: "These persons do not live in accordance with the standards of an inner world regulated by guilt feelings — rather, they live in an archaic world, which, as they experience it, has inflicted the ultimate narcissistic injury on them, i.e., a world that has withdrawn its empathic content from them after having first, as if to tease them, given them a taste of its security and delights. They responded to this injury by becoming super-empathetic with themselves and with their own needs, and then have remained enraged about a world that has tried to take from them something they consider to be rightfully their own — i.e., the parental, empathic self-object." (Kohut, Heinz, *The Search for the Self,* 1978, page 831.)

John Victor Stoen: A Little Boy's Life Lost in Jonestown's Apocalypse

Of the 918 who died at Jonestown, more than 200 were children. One of these children was John Victor Stoen, age six. His father, Tim Stoen, was a Stanford Law School graduate and assistant district attorney of Mendocino County who had joined the People's Temple because of its social commitment and its atmosphere of racial harmony. His mother, Grace, was also a cult member. John Victor, known as John-John, was born in 1972. That February, almost two weeks after his son's birth, Tim Stoen signed an extraordinary document, witnessed and signed by Marceline Jones:

"I, Timothy Oliver Stoen, hereby acknowledge that in April 1971, I entreated my beloved pastor, James W. Jones, to sire a child by my wife, Grace Stoen, who had previously, at my insistence, reluctantly but graciously consented thereto. James Jones agreed to do so, reluctantly, after I explained that I very much wished to raise a child, but was unable, after extensive attempts, to sire one myself. My reason for requesting Jim Jones to do this is that I wanted my child to be fathered, if not by me, by the most compassionate, honest, and courageous human being the world contains." (Quoted by L. Wright, page 75.)

Jones was strict about the sexual behavior of Temple members, but his own infidelities were well known and I think unconsciously strategic as expressions of his narcissistic rage. His followers were expected to make allowances because their leader's psychic gifts were so highly charged with sexual energy and so constantly in need of release that Jones claimed to masturbate 30 times a day. The evolving philosophy of the cult was that everything was held in common – even children, even the bodies of members. Jones, of course, was the controlling Father-God.

Grace Stoen defected in 1976. Tim Stoen left in 1977. Though divorced, Tim and Grace united in a long legal fight and won custody of their son, but Jones would not return the boy. Little John-John represented a loyalty battle in which Jones fiercely enmeshed his whole cult's destiny. When Congressman Ryan began to investigate,

48

Jones called his attorney, Charles Garry, saying that 300 Jonestown residents were ready to commit suicide if any authority came for the boy.

According to Reiterman, Stanley Clayton (a People's Temple member) witnessed the following before he saved his own life by fleeing into the jungle as the White Night of Apocalypse was going on: "He (Clayton) had seen Annie Moore (one of Jones's lieutenant paramours and the babysitter of John Stoen) leaving the pavilion area with John Stoen. The boy had been crying, sniffling, and saying, 'I don't want to die. I don't want to die.' Jones saw him and said, 'Is that my son doing all that crying? My son shouldn't be crying.' John didn't fully stop; he tried to jerk away from Annie Moore as they walked towards Jones's house." (*Raven*, page 561.)

John Victor died from poison ingestion or injection in Jones's house rather than with Jones at the podium of apocalypse.

Jim Jones had literally and figuratively seduced John Victor Stoen's parents, Tim and Grace. Tim Stoen and Grace Stoen had not paid the piper of good parenting. Their dependent narcissistic attachment and worship of Jones in the late sixties and early seventies led them to neglect John Victor and allow him to be taken over by Jones. Jones was obsessed with John, convinced that the boy looked like him and would be the heir to the People's Temple throne.

Days before the Apocalypse of group death, Leo Ryan and his delegation confronted Jones about John Victor Stoen. Reiterman describes Jones's rambling response as it moved toward the topic of John Stoen (*Raven*, pages 496-497) in a free-wheeling interview with Jones by Reiterman and Leo Ryan, with interspersed direct quotes from Jones via recordings made at the time.

"'I brought 1200 people here.' He spoke of himself as a prisoner. He said he needed the medical treatment ordered by Dr. Goodlett, but could not leave Jonestown for fear John Victor would be kidnapped, for fear things would collapse in his absence. 'In some ways, I feel like a dying man,' Jones said. And again the child custody matter surfaced. 'This is very painful,' he said. 'Now he (John Victor) doesn't know who his mother is. He forgets his mother. I haven't taught any hate to the child. [NO, *just vacillation between neglect and overstimulating indulgence by multiple mothering surrogates among Jones's paramours and servile Temple babysitters.*]

"Jones denied outright that John Stoen had been turned systematically against his mother. 'We teach love,' he said, as the six-year-old in a brown and yellow striped T-shirt was brought to his left side. The handsome olive-complexioned child appeared bashful and a little uneasy, yet curious about the TV equipment that suddenly appeared. He squinted into the lights.

"'We have the same teeth and face,' Jones said, baring his own teeth and pulling off his tinted glasses for the first time. Taking the boy's chin between his thumb and fingers, Jones squeezed gently, making John show his teeth for comparison. The child was being treated like a show dog on parade. It was grotesque.

"'He is very bright,' Jones said, stroking and holding him. The passive child projected no warmth.

"'John, do you want to go back to live with Grace?' Jones asked.

"'See?' Jones said approvingly. 'It's not right to play with children's lives.'" (*Raven*, page 497.)

Jim Jones's Dark Epiphanies

Certain events in Jim Jones's life functioned as Dark Epiphanies to propel him toward destruction and confirm his own paranoia and fears. The first may well have been his rejection by Father Divine's congregation and with it Divine's image as a person of power and wealth. The investigations begun by the IRS and the State Department in the 1970s were initially only perfunctory, but Jones magnified them as they fed his own paranoid projections. The defections of cult members before and after the move to Jonestown were taken as severe wounds. This was particularly true of Grace and Tim Stoen's departure. The Stoens's legal efforts to regain custody of young John Stoen seemed to threaten Jones's narcissistic core. The visit by Congressman Ryan's delegation seemed to represent the culminating Dark Epiphany that made the apocalypse loom inevitable.

Jones's Apocalyptic Plan

In apocalyptic cults, a key and unique element is woven into the core of cult ideology by the charismatic leader. The leader repeatedly holds forth the ideal and noble fantasy of dying together for the grandiose, if not sacred, cause. The powerful ideological "ace in the hole" of apocalyptic group cult life is for everyone to be prepared to die in a mass suicidal act to protest the evils or injustice of the outside world as declared and dramatized by the leader.

The group death scenario is not a sudden, impulsive commandment instigated by the leader at a time of crisis, but is woven into the ongoing ideology and day-to-day theology of the group. In California five years before the apocalypse and later in the Guyana jungle, Jones habitually kept the members awake long into the night (after their many hours of grinding labor) with "white-night" exercises in which he prepared them for ultimate destruction. Audiotapes of these free-associative sermons reveal his descriptions of potential sacred battles against threatening forces. Jones even brought out vats of liquid for practice runs, forcing and encouraging members to drink the harmless potion, so that by the time of the November 1978 suicide, the members had been so desensitized to the act of suicide that mothers poured the poison into the mouths of their own children. Thus the suicide scenario was taught, conditioned, indeed brainwashed into the resonating group identity of the cult.

Foreshadowing Suicide

On Memorial Day in 1977, a year and a half before the Jonestown disaster, Jones and 600 of his followers showed up at the Golden Gate Bridge in San Francisco at an anti-suicide rally. Each cult member wore an armband bearing the name of one of the bridge's victims. With local news media looking on, Jones delivered a general appeal of concern for suicide victims. Then abruptly his tone and demeanor changed; suddenly *he* was the one being driven to the brink. "Suicide is a symptom of an uncaring society," he said. "The suicide is the victim of conditions which WE cannot tolerate, and, and ..." He paused, realizing that he had misspoken. "I guess that was Freudian," he commented, "because I meant to say, 'which *he* cannot tolerate,' which overwhelm him, for which there is no recourse."

Then Jones became even more direct. "I have been in a suicidal mood myself today for perhaps the first time in my life, so I have personal empathy for what we are doing here today." (*Raven,* page 321.) Although I believe Jones's predominant diagnosis was Narcissistic Personality Disorder, which in time worsened towards Malignant Narcissism and eventually psychosis, it is possible that he was also depressed. He also was a substance abuser and may have had a central nervous system infection. (See "Diagnosis," below.)

Jones as a Father-Seeker

Like other destructive and exploitive cult leaders, Jones began a relentless quest to become a strong parental figure for other people. Like Father Divine, he insisted that his followers call him Father or Dad. For a Malignant Pied Piper, acting as a parent to cult members finally allows an ephemeral experience of a good parent within themselves. This life-long search requires a ready supply of child-admirers. They give psychological birth to these followers through the recruitment processes discussed in Chapter Six. In essence, Jones and other cult leaders gain a sense of power over and mastery of their own childhood psychological deformities by becoming overwhelmingly significant and powerful in the lives of their followers.

Yet beneath his outward confidence and swagger was a profound, unconscious sense of shame, humiliation, narcissistic rage, and fear of aloneness. Rage was ready to surface whenever his progressively fragile narcissism was threatened or punctured. Thus the cult family honeymoon was eventually over and the formerly abused, neglected, and abandoned leader became the all-powerful abuser, neglecter, and abandoner.

Freud's imaginary soliloquy about the psychological reaction of Shakespeare's Richard III to his congenital physical deformity is applicable here: "I have the right to be an exception, to disregard the scruples by which others allow themselves to be held back. I may do wrong myself, since wrong has been done to me." (Freud, 1916.) This extreme sense of entitlement led Jones to lure more than 900 people to the steamy Guyana jungle. There he finally found his father-himself and a cult family to surround and embrace him in the group death. Not alone, yet so alone at last.

An Obsession with History
Jones was obsessed with his legacy and spurious place in history. Over time, he presented himself as Christlike, then as Godlike; and eventually as a Marxist/socialist God who was an embodiment of Lenin. He became his own grandiose Father-God himself. Some of his other heroes were Patrice Lumumba, Stephen Biko, Salvadore Allende, Paul Robeson, and Victor Jara (the last a Chilean poet whose songs uplifted the poor during Allende's ascent to power, and who was tortured and martyred).

Shortly before Jones died, he called out, "Jehovah Jara, I'm joining you." (quoted by James Reston, Jr., in *Our Father Who Art in Hell,* page 337.)

Reston also describes Jones's wooden aquamarine-color throne of death with armrests and mussed pillow. It was raised on a white pedestal. Above Jones's slumped dead body was a sign in creamy block letters on a dark background: "Those who do not remember the past are condemned to repeat it."

Reston points out that the epigram was plucked from philosopher George Santayana's book *The Life of Reason,* and comes from a passage with this context: "Progress comes in manhood and rests on the ability to change, to adapt, to retain past experience and learn from it. When past experience is not retained, infancy is perpetual — the condition of children, savages, and barbarians." (Reston, page 336.) Reston also says that Santayana probably would have been amused at how his epigram, so glibly quoted by Jones, turned into Jones's epitaph. In the same passage, Santayana also wrote, "It is remarkable how inane and unimaginative Utopias have generally been." (Reston, page 337.)

Jim Jones's Rage

Jones's cult members called him Father, at his behest, and he loved to boast of their loving merger. At the same time, he exploited and devalued them just as he felt he had been exploited and devalued as a child. A portion of a tape from the Guyana jungle, recorded 11 months before Congressman Ryan's ill-fated visit, shows Jones's hatred and loathing for his followers. (Reston, page 223 ff).

"Oh, we've got different distinguished people who want to visit us, to investigate us. I've got all kinds of programs for them, and they're worked out from A to Z." [cheers] "You're so naïve. You don't even know what Jim Jones is all about. You can't even follow him. You haven't even smelled where he is at yet, much less follow him." [shouts for more]

"You are stupid pissants and reptiles, who are lower than the primates, you can make whoopee if you want, but your whoopee makes me sickly ... peace, peace ... You make your whoopee, while I do something that's far more significant, because I know exactly what's going to take place. I've made some big plans, honey ...

"You fuckers, I like to look at you now, because you don't know how clever I am. I made plans for your treason long ago, because I knew I couldn't trust nothing, only Communism, and the principle that is in me — that *is* me!"

Jones paradoxically enticed his followers with virtuous ideology and personal charisma, and then humiliated and devalued them into psychological serfdom.

According to Carl Goldberg (*Psychodynamic Perspectives on Religion, Sect, and Cult,* Halperin, ed., 1963, page 172), Jones was a genius at "stripping the defenses of his parishioners at all-night 'catharsis' sessions. During these evenings, members would be accused of offenses, humiliated, and beaten." Jones combined psychedelic and mind-altering drugs, deficient diets, and sensory or sleep deprivation to draw his followers into induced, yet subtly sought-after, dependent states. Although the Jonestown members had ready evidence of Jones's hypocrisy, instability, and cruelty, they went to extreme lengths to deny what they saw and thought. In

Jones's jungle kingdom, fear ruled, and contradiction, poverty, and continual labor were the spurious courtiers.

Diagnosis: Malignant Pied Piper

Jim Jones fits the criteria for Narcissistic Personality Disorder, with the additional elements of Malignant Narcissism.

Narcissistic Personality Disorder (five or more of the following). Jim Jones exhibited seven or eight out of nine criteria.

1. Has a grandiose sense of self-importance: On the surface Jones championed racial integration and projects to provide food and shelter for the poor and elderly, but "…under that was a personal need to be admired, loved, and lauded by the crowds." (*Raven,* page 44.) Steadily and relentlessly Jones began to insert the idea that he was God.

2. Is preoccupied with fantasies of unlimited success, power, brilliance, beauty, or ideal love. Reiterman captures this issue about Jones well: "Subtly he encouraged his rank and file to see him in Christlike and Godlike terms. Though his aides complained privately, Jones rationalized that his people needed the false illusion of his deification in order to dedicate themselves totally to the Temple's worthwhile goals. Only decades later — in Jonestown, when he had an unbreakable hold on hundreds of people — would he drop most pretenses of being God himself.

"Jones's assistants, in Indiana and later, never challenged the self-aggrandizing tendencies of their leader. By not publicly opposing him, they gave tacit endorsement." This last point has profound implications for avoiding the spiritual clutches of an MPP. Individuals, family, parents, friends need to actively challenge and confront the "Emperor's new clothes" without being vindictive or insulting.

3. Believes that he or she is "special" and unique and can only be understood by, or should associate with, other special or high status people or institutions. Jones approached and received the approval of Rosalyn Carter, Walter Mondale, and San Francisco mayor Moscone, among others. He visited Russia and Cuba to court favor with Castro and other communist leaders. (Reston, Chapter 7.)

4. Requires excessive admiration. Jones had endless and insatiable varieties of maneuvers, preaching/healing techniques, cons and manipulations to get personal admiration, sexual affirmation, and devotion of his followers. By attacking and devaluing his followers' families, countries, and communities of origin, he cemented their admiration for himself. (Reston, page 269.)

5. Has a sense of entitlement, i.e., unreasonable expectations of especially favorable treatment or automatic compliance with his or her expectations. Jones reacted with woundedness or haughty arrogance when criticized or rebuffed. See the account by Reston (page 181) of Jones's reactions to negotiations about possible People's Temple exodus to Russia or Cuba. Jones seemed to expect exact compliance by Russian authorities.

6. Is interpersonally exploitive, i.e., takes advantage of others to achieve his own ends. How Jones elicited "testimonials" from his followers at the long and frequent congregational meetings confirms this point. Reston gives an example of his maneuvering of an elderly woman at one such meeting (Reston, pages 183-184 and 223). Jones whips up the congregation to support a hunger strike if he doesn't get access to their Social Security checks for his cult coffers.

7. Lacks empathy; is unwilling to recognize or identify with the feelings and needs of others. Most of the MPPs in my study do not fit this criterion exactly. Jones recognized and tuned in exquisitely to his recruits' and followers' feelings and needs, but it was predominantly so he could in turn control them and insure their loyalty, obedience, and adoration.

8. Is often envious of others or believes that others are envious of him. The grandiosity of all MPPs is so great they seem to assume they are superior to even being envied.

9. Shows arrogant, haughty behaviors or attitudes. On numerous occasions after verbally seducing and exciting his congregation, Jones would suddenly turn on his flock, devaluing them with haughty and arrogant language and asserting his superiority to them.

Malignant Narcissism (four out of four criteria).

1. Paranoid regressive tendencies with brief lapses in reality testing and rational thinking. Jones showed this verbal behavior while preaching to his followers or remonstrating with his henchmen. (Reston, page 267.)

2. Chronic self-destructiveness or suicidal behavior as a triumph over authority figures. Jones wove this personal dynamic into the group life of the People's Temple via the "white night" group suicide rehearsals that went on for years before the final apocalypse in Jonestown. (*Raven*, pages 294-295.)

3. Major and minor dishonesty (psychopathy). Jones would fake cancer healings at temple services, and secretly

obtain medical information about people to set them up at healing services and other chicanery. (See *Raven,* pages 44 and 51.) Jones was under investigation for Social Security violations; inappropriate taking of members' property (Reston, pages 299-300); and even beatings, blackmail, and forced confessions of followers. (See Colin Wilson's *Rogue Messiahs,* page 51.)

4. Malignant grandiosity with sadistic efforts to triumph over all authority. Kernberg's observations tune in to destructive cult leaders like Jones in an uncanny way. Kernberg implies that these persons' grandiosity and self-pedestalizing are reinforced by the sense of power and triumph they achieve over their underlying fear and pain through inflicting pain on others. This process brings them a sexualized "joyful cruelty" that allows them to obtain a sense of superiority and triumph over life and death via their own process of suffering and that of their followers. (In MPPs, this triumph resembles, in my opinion, a satisfying murder of a hated father who abandoned his child without remorse, and now the tables are turned!) Jones and Manson most vividly exemplify this fourth element of malignant narcissism. Jones even sodomized male Temple members in front of their wives to "help" them accept their latent homosexuality and affirm Jones's superiority! (*Raven,* page 175 and 177.)

In the final months and weeks in Jonestown, it was clear that depression (probably major depression with psychosis) came into the final diagnostic picture. The "white nights" and revolutionary group-suicide plans and rehearsals had gone on for many years and I think were part of Jones's malignant narcissistic and psychopathic personality dynamics.

In the later phases of Jones's personality disorganization, substance abuse was also a factor. There are also some reports that

Jones may have suffered from a fungal infection of his brain (disseminated Coccidiomycosis) contracted in the jungle months before the November 1978 apocalypse. This was not confirmed at autopsy. (*Medical World News,* Dec. 11, 1978.)

Lynetta Jones's Poems
Written by Jim Jones's mother, Lynetta, a "momma dearest" of the first rank. (*Raven,* pages 7 and 207.)

(untitled)
I took a piece of plastic clay
And idly fashioned it one day,
And as my fingers pressed it still,
It molded, yielding to my will.

I came again when days were past,
The bit of clay was firm at last,
The form I gave it, still it wore,
And I could change that form no more.

A far more precious thing than clay,
I gently shaped from day to day,
And molded with my fumbling art,
A young child's soft and yielding heart.

I came again when years were gone,
And it was a man I looked upon,
Who such godlike nature bore
That men could change it – Nevermore.

Ode to Liars
Wherever you have lied,
Permanently or to stop awhile,
You will make your little lies come forth,
Little thin wisps of lies at first,

Then larger windblown ones,
Complete with all the anger
That seethes within yourself ...

An Absence of Trust

Famous psychoanalyst Erik Erikson wrote about the development of basic trust as part of the child/mother relationship: "The amount of trust derived from the earliest childhood experience does not seem to depend on absolute quantities of food or demonstrations of love, but rather on the quality of the maternal relationship. Mothers create a sense of trust in their children by that kind of administration which in its quality combines sensitive care of the baby's individual needs and a firm sense of personal trustworthiness within the trusted framework of their culture's lifestyle. This forms the basis in the child for a sense of identity which will later combine a sense of being 'all right,' of being oneself, and of becoming what other people trust one will become ... But even under the most favorable of circumstances, this stage (Basic Trust) seems to induce in psychic life a sense of inner division and universal nostalgia for a paradise forfeited" (Erikson, 1950).

We know that beyond her sanctimoniousness and grandiose ambitions for her son, Lynetta Jones was not there for him at the level of trust described by Erikson. Mother Jones could have chosen to leave with Jim and not just put his father down. She could have helped young Jim to understand his dad's medical situation. She might have obtained reliable childcare. Without solid Basic Trust, it is no wonder that Jones had a flawed sexual and personal identity. Rather than a sense of generativity (concern for the identity, security, and welfare of the next generation), Jones stagnated as a person, and in bitter, paranoid despair he took his fantasized replacement, young John Victor Stoen, to a group death with him. Rather than caring for the next generation of the People's Temple, Jones destroyed it.

Like Mother, Like Son

Clearly, Lynetta Jones had a dominant influence on the development of her son's character disorder. Although she spent little personal time with her son, whose overriding memory of his childhood was one of anger and loneliness, she added the pressure of vicarious involvement and seduction. Her devaluation of his father and her overweening ambitions for her son only encouraged him to act out her forbidden wishes and angry disappointment and rage. To her, the boy became a symbol of her own ambitions, and in the Guyana jungle she believed that her destiny would merge with his. She also instilled in Jim her own scorn for psychiatry and its effectiveness.

In 1952 doctors Adelaide Johnson and Stanley Szurek wrote a classic paper called, "The Genesis of Antisocial Acting Out in Children and Adults" (*The Psychoanalytic Quarterly,* Vol. 21, pp 323 ff). Johnson was a child analyst who saw a delinquent child or teenager in psychotherapy. Szurek was an adult analyst who saw the child's parent(s) in analysis or psychotherapy. With their patients' permission, they shared and compared the parallel therapies. Basically they concluded that the overt areas of acting-out (antisocial) behavior in the child were the same areas of unconscious conflict, preoccupation, and fascination of their parent(s). The parents participated vigorously and vicariously.

This seems to hold true for Lynetta Jones and her son. Reiterman says this about Lynetta's influence: "Along with her canny aggressiveness, she passed on to the child a general irreverence toward the world; a certain self-righteousness and grandiosity that allowed him to see himself as an independent spirit proudly pushing against the prevailing flows of society. Along with a certain sensitivity, she gave him a mission — a sense of wrongs to be corrected, a feeling of persecution, a resolve to fight back for his mother and himself." (*Raven,* page 160.) Yet she also left him alone and without a close family or parental coalition to guide him. His own inner anger and rage at his childhood deformities may have been expressed in his obsession with nuclear holocaust (a powerful symbol of dramatic and catastrophic destruction that mirrored his own explosiveness) and in his flight to Guyana as a futile attempt to escape.

When Lynetta Jones died in Jonestown in December 1977, one year before the People's Temple apocalypse, Jones was devastated. He had lost the source of his power and his sickness, and at the same time had lost one of the few feisty restraints on him. His life and that of more than 900 followers moved inexorably toward doom.

Fifteen years after the Jonestown debacle, another Malignant Pied Piper named David Koresh led 90 followers to a violent death in Waco, Texas. Like Jones, Koresh had a lonely, narcissistically injured, and empathy-starved childhood. Like Jones, Koresh reacted with extremes of entitlement and rage acted out upon and with his cult group.

CHAPTER THREE

DAVID KORESH:
MALIGNANT PIED PIPER OF WACO, TEXAS

"My hand made heaven and earth, My hand also shall bring it to the end ...Your sins are more than you can bear. Show mercy and kindness and you shall receive mercy and kindness ... You have a chance to learn My Salvation. Do not find yourselves to be fighting against Me ... Please listen and show mercy and learn of the marriage of the Lamb. Why will you be lost?"
—Yahweh Koresh, April 11, 1993

The Critical Incident at the Branch Davidian Compound

On February 28, 1993, Federal agents tried to execute an arrest warrant for David Koresh as part of an investigation into allegations of illegal weapons and child abuse at his Branch Davidian cult's Mount Carmel compound ten miles from Waco, Texas. A shoot-out between the government agents and the cult members left ten people dead: four agents from the Bureau of Alcohol, Tobacco, and Firearms (BATF) and six Branch Davidian sect members.

On April 19, 1993, Koresh and more than 85 followers died in a fiery group murder-suicide event at Mount Carmel. The apocalypse occurred after confused negotiations by Koresh with U.S. government authorities during a 51-day siege.

Twenty-five children died during the apocalypse. Many of them were Koresh's own children by various female cult members. Twenty-one children survived the fire because Koresh released them five days after the initial shoot-out.

The Child Survivors at Waco

Dr. Bruce D. Perry, chief of psychiatry at Texas Children's Hospital at the time, led a team that worked for two months with the surviving children of the Waco apocalypse. Perry's team interviewed 19 of the 21 children who had been released (a three-year-old toddler and a seven-month-old infant were too young to be interviewed). The children revealed that David Koresh told them to call their parents "dogs"; only he was to be referred to as their father. Girls as young as 11 were given a plastic Star of David, signifying that they had "the light" and were ready to have sex with the cult leader. For sins as small as spilling milk, the children said, they were struck with a wooden paddle known as "the Helper." To train for the final battle, they were instructed to fight each other, and if they did not fight hard enough, they were paddled for that, too. *(The New York Times,* May 4, 1993, page 1.)

Dr. Perry described the children's world as "a misguided paramilitary community" in which sex, violence, fear, love, and religion were all intertwined. In an arrangement typical of destructive cults controlled by Malignant Pied Pipers, Koresh regulated the sex lives of members. He ordered adult men and women to be strictly segregated in the compound. But the children told Dr. Perry that Koresh had "wives" as young as 11 years old and he routinely discussed sex openly with even the youngest girls in Bible lessons.

Dr. Perry described the children's experiences in the cult world as "pseudonormal." They drew hearts with "I love David" scrawled in the middle. Dr. Perry concluded that the children were conditioned to substitute the experience of Koresh's version of "love" for fear. Dr. Perry said that the cult leader controlled everything in the children's lives — sex, school, play, and diet. He became the ultimate controlling parent-himself. During the children's therapy sessions with Dr. Perry, they drew pictures of fires, explosions, and castles in heaven. Dr. Perry wrote in his report about "the sense that there is going to be an absolute explosive end to these children's families." He conveyed these findings to the FBI authorities in written reports during the siege. *(The New York Times,* May 4, 1993, page 13.)

The Life of David Koresh
David Koresh was born Vernon Wayne Howell in Houston, Texas, on August 17, 1959. He was not yet 34 years old when he died at Waco. For the purposes of this discussion, we will refer to him as David Koresh except in remarks about his childhood. He started calling himself David Koresh in the 1980s and changed his name legally in 1990.

Koresh's mother, Bonnie, was 14 (the age of some of Koresh's "cult wives") when she became pregnant by her 20-year-old boyfriend, a carpenter named Bobby Howell. The parents never married and two years after the boy's birth, his father fell in love with another woman and left for good. The boy rarely saw his father in the ensuing years and never told stories about him except to say he did not like him.

Koresh later said on a taped sermon, "I was only born because my daddy felt something in his loins and lusted after my momma." (Breault and King, *Inside the Cult,* page 27.)

Young Vernon was cared for by his mother's sister and his maternal grandmother. He thought his aunt was his mother, and did not know that his "Aunt Bonnie" was his biological mother until he was five years old. Then Bonnie married and the boy was told that "Aunt Bonnie" was his real mother and he would live with her. In a sermon years later to the Branch Davidian sect, Koresh said, "I was shocked! I was confused — here I was five years old and my whole world was turning upside down." (Breault and King, page 28.) Young Vernon didn't get along with his stepfather and was soon shuffled back and forth between relatives until he dropped out of school in the ninth grade because he was a poor student. Other kids teased him and called him "Vernie." He was dyslexic and was troubled by compulsive masturbation as an adolescent. But he did have musical ability and a ravenous interest in the Bible. Vernon memorized huge portions of scripture, including the entire New Testament, and was an active church member. He also excelled at track, and lifted weights to protect himself against physical abuse by his step-siblings and cousins. (David Thibodeau with Leon Whiteson, *A Place Called Waco,* page 39.)

As a young man, David Koresh never got along with employers — he held a series of odd jobs but didn't last long at any of them. Breault summarizes, "Vernon (Koresh) got his bosses' hackles up with his arrogance and holier-than-thou attitude. More often than not, it was Vernon who walked out, rather than take orders from those who he saw as his inferiors." (Breault, page 33.) Koresh also opposed all authority figures in the Southern Baptist Church, Seventh Day Adventist Church, or the Branch Davidian sect, an offshoot of the Adventists.

In the 1980s, he chose the name David Koresh. "David" was his connection to the Biblical King David; "Koresh" is Hebrew for Cyrus, the Babylonian king who allowed the Jews to return to Israel. (*Newsweek,* March 15, 1993, page 57.) Vernon Wayne Howell's new name reflected his grandiose identification with God and the prophets of the Hebrew Bible. Koresh also believed he was the seventh and last angel mentioned in Revelation 10:7, the agent of God who would bring about the end of the world.

The Roots of David Koresh's Psychological Deformity

Koresh, like Jim Jones, had an absent, weak, and abandoning father in his childhood. Koresh, like Jones, tried over time to become the powerful father/himself in his cult. Koresh called his gun-proficient cult lieutenants his "Mighty Men." The machismo reverberated back and forth between them. Koresh and Jones both enacted within their pitiful leadership efforts the "strengths" they felt they had missed out on in their childhoods from their fathers.

In his late-night dialogues with FBI agents during the siege, Koresh described his childhood as lonely. *(Frontline,* "Waco: The

Inside Story," March 1993.) In a film interview in the presence of his followers, Koresh, then in his thirties, had the following exchange with Martin King about "the old days" of his childhood.

Koresh: "I grew up. I failed school. I quit in ninth grade because I had other … things to do."

King: "Like what?"

Koresh: "I had to learn some things. I'm a student."

King: "What did you learn?"

Koresh: "About people," and he gestured to his followers. "That's what I do, I … learn about people, about myself. [A typical merger of a Malignant Pied Piper with his followers.] These people, the majority of them were college students, they were raised by mom and daddy, they learned what mom and dad had to show them, they rebelled because they had their own minds as children."

Koresh's knowing remark about his followers' rebellion is an example of *projective identification,* which is typical of all MPPs I have studied. *Projective identification* is a term introduced by psychoanalyst Melanie Klein. It is the unconscious mental process of projecting one or more parts of oneself into another person. The projected psychological part may be an intolerable, painful, or dangerous aspect of the self.

In Koresh's case, this meant projecting onto cult members his own rage and rebellion at his parents/himself, and his attendant disdain for all authority except his own. Severely "borderline" or narcissistic people use this complex unconscious mechanism to reduce their own inner anxiety and inner disorganization by projecting them onto another person or persons. When the projector, such as Koresh, does this, he can see the anxiety issue "out there" in the other person; his own anxiety can be assuaged and he can comfort, cajole, or condescend to the others.

It may also be that by identifying himself with college students who had been raised by both parents and given a good education, and then had rebelled anyway, Koresh was trying to deny his own deprived childhood and protect himself from acknowledging its impact on his personality.

A valued aspect of the self may also be projected into the other person for "safekeeping." Just like Jim Jones, David Koresh could at times act tenderly, especially toward children. For example, Waco survivor David Thibodeau described how children participated in every aspect of the cult, including the long study sessions. "Often, a child would sit in David's lap while he was expounding Scripture, cuddling close while he stroked the kid's hair or kissed his or her cheek. He was very touchy-feely with all the children, perhaps in reaction to his own lonely, unloved childhood. Sometimes, amid a fire-and-brimstone exposition, amid talk of lion-headed horses with snakes for tails, he'd pass around a bowl of popcorn to settle the young ones." (Thibodeau, page 117.)

Given Koresh's physical and sexual abuse of the children in the cult, this public tenderness and nurturing seems all the more false and revolting. At the very least, I believe Koresh was projecting tenderness onto the children so that he could enjoy the "good" feeling at a safe emotional distance and not acknowledge his own neediness or longing for it. Malignant Pied Pipers and their followers are a virtual symphony of reverberating projective identifications! *(American Psychiatric Glossary,* page 108.)

Koresh's comments about his followers/himself continued, "Some went into drugs and some went into partying and some into athletics — and finally they came to religion, and they heard, they got an ear load, and they gave those institutions time to show them this great God and this great salvation in Christ."

King: "But you said *you* failed."

Koresh: "I've equal opportunity also to show them something."

King: "What will you show them?"

Koresh: "Well, let me tell you this. I have more knowledge in my little toe than all the great scholars could learn in a lifetime." (King, page 32.) Koresh's grandiose self leaps forth here!

Koresh Finds His God/Himself
Colin Wilson reports that when Koresh was 19 he had an affair with a 16-year-old girl who bore his child. Koresh wanted to marry her, but she felt that he was unfit to raise a child and left him. He

experienced this as a narcissistic wound. (This event was probably an adolescent psychological deformity as well as a contributor to a dark epiphany.)

Koresh then sought solace in the religion of his early childhood, and became a born-again Christian in the Southern Baptist Church. He sought help from his pastor about his compulsive masturbation but the pastor only exhorted David to pray for strength. When this remedy didn't work, Koresh decided that the Southern Baptists lacked any true connection with God.

In 1977, Koresh made an unsuccessful journey to Hollywood to find movie roles as a rock-and-roll musician, and then wandered to many places doing odd jobs. In 1979, he was baptized into the Seventh-Day Adventist Church in Tyler, Texas, returning to the Seventh-Day Adventist religion of his mother. Koresh soon fell in love with the daughter of his new pastor and declared to the pastor that God had given the girl to him by divine revelation. The shocked cleric threw Koresh out, and when David persisted, he was dismissed from fellowship by the congregation, with the agreement of the young woman. (Wilson, Colin, *Rogue Messiahs*, page 10-11.) Other reasons for his dismissal from the church in 1981 included his obnoxious behavior during church services and his constant references to sex in his theological talks.

Later that same year, Koresh found his way to Mount Carmel, where he joined the Branch Davidians as a dishwasher. The Branch Davidians traced their religious heritage back to the 1830s, a time of great religious ferment and sectarian change. In 1831, an atheist-turned-Baptist minister named William Miller began studying the end-time prophecies in the Bible and declared that Doomsday would occur on a particular day in 1843. When it failed to happen, the Millerites dispersed, but their millennial fervor did not, and their sect eventually became the Seventh-Day Adventist Church.

In about 1930, an Adventist preacher named Victor Houteff was dismissed from fellowship in his church in Los Angeles. Houteff founded his own sect, known popularly as The Shepherd's Rod, which predicted a kingdom in Israel ruled by Jesus Christ and his lieutenant, "Antitypical David."

When Houteff died in 1955, his widow, Florence, predicted that 1959 would bring a literal slaughter of the wicked Seventh-Day Adventists and the resurrection of her husband. Another Adventist, Ben Roden, disagreed, saying it would happen in 1960. Roden split from Houteff and founded his own movement, called the Branch Davidian Seventh-Day Adventists. At his death in 1979, his wife Lois assumed the presidency of the Branch Davidians.

In 1981, when Koresh arrived at the Branch Davidian compound in Mount Carmel, he was instantly disliked for his arrogance. Colin Wilson describes Koresh's early boundary-trampling maneuvers there: "Lois Roden, the head of the sect, was everything [Koresh] had dreamed about. Still attractive at 68, she was a famous TV evangelist, a friend of the rich and famous, who spent much of her time traveling around the world. She was also a favorite of the feminist movement, since she had announced that God was female, and began the Lord's Prayer 'Our Mother, who art in heaven.'

"For a long time, she shared the general view of the new recruit, and made [Koresh] live in a small, unfurnished room to try to cure his conceit. Her view began to change when, two years after his arrival, he told her that the Lord had revealed to him that he had been chosen to father her child, who would be the Chosen One. When Roden's son, George, who expected to replace his mother as president, found out, he did his best to eject the interloper. His mother, convinced she was pregnant, defended [Koresh]. The power struggle ended abruptly when [Koresh] announced that God had ordered him to marry a 14-year-old named Rachel Jones." (Wilson, page 11.)

In his seduction of Lois Roden and subsequent abandonment of her, then his seduction of a young girl the same age his mother had been when she bore him, Koresh acted out his own unconscious narcissistic rage against women. A recurrent theme in his life is his pattern of seduction and abandonment of women, just has he had once felt abandoned and hurt as a boy.

For a while Koresh and George Roden had a fragile truce. Then, in a rage, George fired his Uzi at Koresh. He was a bad shot, but Koresh departed for the "wilderness" with a splinter group of 25 followers. They settled in Palestine, Texas. While his followers roughed it, Koresh traveled to California, Israel, and Australia on recruiting missions. Koresh was literally tossed out of a Seventh-Day Adventist Church in San Diego when he got up during a service to announce that he was the Messiah. (Wilson, page 12.) Yet Koresh's amazing, encyclopedic knowledge of scripture and ability to preach convincingly about it progressed. The group in Palestine grew steadily. If a new disciple had a teenage daughter, she usually became Koresh's "wife."

Meanwhile, back in Waco, Lois Roden died. Her son, George, grew paranoid and ruminated jealously about Koresh's success 90 miles away in Palestine. George Roden declared himself to be God, and in 1987, he made a bizarre challenge to Koresh. He dug up the body of an 85-year-old deceased member of the Branch Davidian congregation and challenged Koresh to a contest. Whichever of them could raise her from the dead would be the true prophet of God.

Koresh declined the challenge but pressed charges on George for abusing a corpse. The police said they needed pictures of the corpse, so Koresh and some of his armed "Mighty Men" tried to sneak into Mount Carmel for the pictures on November 3, 1987. Roden caught them and a gun battle ensued. None of the combatants was mighty enough to hit anyone, and the police arrested everyone.

Out on bail, Koresh told his side of the situation to a local television station. Roden got so enraged that he wrote letters to the Texas Supreme Court threatening to strike down everyone with AIDS and herpes if they did not jail Koresh. George got six months in jail for contempt and a jury acquitted Koresh and his "Mighty Men."

Fifteen months later, when a Branch Davidian follower suddenly declared Koresh to be the Messiah in Roden's presence, Roden murdered the man by splitting his head open with an axe. Roden was convicted and put in prison. The Mount Carmel compound had many debts, but Koresh and his followers raised the money to purchase it, and the Palestine group merged with the Mount Carmel Davidians.

Koresh had banished his rival and his Oedipal victory was complete, but he did not live happily ever after.

Koresh's Dark Epiphanies

Koresh's search for father figures was repeatedly frustrated by his unresolved inner rage at being rejected and abandoned by his father and stepfather. Koresh set up his encounters with male authority figures at jobs and in churches to insure repeated rejection experiences. (The ultimate father-authority confrontations would occur with the FBI and ATF.) Lacking real fathering, his path continually led him to be a grandiose father within himself. He did this by projecting the illusion onto his cult followers that he was the ultimately powerful father. This illusory sense of restitution and revenge is found in all Malignant Pied Pipers.

Koresh also felt abandoned and hurt by women early in his life. He unconsciously acted out this rage at women by sexually and psychologically exploiting, controlling, abandoning, and hurting them. Starting when he was 16, he had a long string of sexual affairs, including his supposed impregnation of Lois Roden, at age 68 clearly a grandmother figure, whom Koresh threw over for Rachel, a 14-year-old he impregnated and married. (Remember, Koresh was passed back and forth between grandmothers' tenuous care as a boy.) He later married and impregnated Rachel's 40-year-old mother, and seduced or raped many other female cult members.

Despite his lofty preaching and encyclopedic knowledge of scripture, he steadily violated ethical and moral principles. He had a propensity for dishonesty, blurred sexual boundaries, impulsivity, and antisocial acting out. His sense of narcissistic entitlement was enormous. In his inner self, David Koresh was probably that lonely, hurt, and abandoned little boy Vernon who felt that he was now above the usual sexual and moral values of society. To get an inner sense of restitution, revenge, and triumph over his parents, who only had "lust in their loins" and no love in their hearts for him, little "Vernie" showed them all.

How a Good Father Provides for a Boy and Young Man

In contrast to the fatherly absence, abandonment, and neglect that Malignant Pied Pipers experience are the benefits that a psychologically healthy and loving father gives to his son. In psychoanalytic terms, a young male needs an adult male role model to identify with at both conscious and unconscious levels. A man's sexual identity and gender-specific social, vocational, marital, and parenting behaviors depend on the loving presence of a male to at first imitate, and with whom he can gradually identify.

If the relationship between a boy and a father or father surrogate is solid, the young man will learn how to develop a healthy relationship to authority and authority figures. He will then assert his own authority in effective and appropriate ways. The inner confidence in one's own authority allows a young man to be assertive but not destructive or exploitive like David Koresh or other MPPs.

The young man with a healthy father can learn to treat a mother, sister, girlfriend, and wife with loving, caring, and tender respectfulness. These fortunate young men will grasp how a mature man or woman can disagree and argue without losing control, bullying, or becoming cruel, violent, or selfish.

Koresh and all the MPPs I have studied did not have the benefit of an experience with a good father or surrogate father. They all had lonely childhoods with parental separations, intense experiences of rejection, neglect, disappointment, shame, and loss. They all had weak, absent, or totally unimpressive fathers. These childhood psychological deformities formed the soil and roots for the accumulating rage resulting from later Dark Epiphanies.

Can a Good Mother Overcome a Defective Father?
In my life and my clinical work I have known excellent strong and loving women who were single parents and successfully raised confident and competent sons by themselves. They made sure that their sons had many healthy men as role models at church, in school, or in the community. I shared an office suite with one of these women and admire her greatly.

The problem with the MPPs profiled in this book is that their mothers are often weak figures as well – or if they are interested in their sons, as Lynetta Jones was, it is in an unhealthy way. David Koresh's mother was so young and lacking in direction that she had precious little to offer as an effective adult role model or empathic presence.

Koresh's Apocalyptic Scenario

> *"My hand made heaven and earth, My hand also shall bring it to the end ... Your sins are more than you can bear."*
> —part of Koresh's final written message before the apocalypse at Mount Carmel

After he became "king" at Mount Carmel, David Koresh was not content. He never seemed satisfied no matter how many new cult recruits adored him or listened in rapt attention to his sermons. Former follower Marc Breault, who sometimes played bass in the rock band Koresh organized at the compound, said that even practicing together was difficult. Koresh threw tantrums when he hit a wrong note on his guitar in front of others. Breault said at an interview, "It's very difficult being in a band with God's messenger." *(Time,* Special Report, May 3, 1993, page 35.)

Because his theology and preaching were tied inexorably to the core of Koresh's pathological narcissism and psychopathic personality dynamics, he progressively made self-fulfilling prophesies about the end of the world. This dynamic will also be noted with Asahara, who orchestrated the Tokyo gas attacks to fulfill his own writings and prophesies about attacks upon Japan by the West.

Koresh played his Pied Piper music on his guitar and with the seductive sermons he preached to attract followers. The death by fire at Waco was precipitated by Koresh's illegal activities. The FBI and ATF agents unwittingly played into the hands of Koresh by their actions, which appeared to fulfill Koresh's idiosyncratic Biblical prophecies. Koresh veiled his delusions of being Christ, King David, or a special prophet by his interpretations of the "Seven Seals" in the Book of Revelation. His prophecies of his cult's suicidal-homicidal death by fire and bullets became self-fulfilling.

As God's self-appointed agent, Koresh initially predicted that the end would come when he moved to Israel and converted the Jews, triggering a war that would cause American troops to invade the Holy Land and signal the beginning of Armegeddon. He actually went to Israel in the 1980s, but when his prediction foundered, he changed his prophecy. After changing his name in 1990, he foretold that the final Armegeddon with the American army would occur in Texas. How vivid must the group-death vision have been when the FBI tanks appeared on the final day in Waco!

Because Koresh was predicting the fiery end of the world, early on he began to prepare and indoctrinate his followers with his apocalyptic scenario. Witness this excerpt from a 1989 Bible study tape: "You stupid idiots! Get it in your minds! He says His heart has declared a teacher! If you don't follow the truth you're going to hell! Psalms 90! You'd better start fearing God [me?], 'cause He's going to burn you in the lowest hell! He's trying to show you He's going to kill you if you don't listen!" (Breault and Martin King, page 123.)

The end times Koresh was predicting also allowed him to expand his fascination with guns into the stockpiling of ever more powerful (and illegal) weapons on the compound. The dutiful followers paid for these with their credit cards, flimsily rationalizing the purchases as good investments. (Only in America!)

"Listen to the Lord! You don't know his fear now! You don't know his terror yet! You haven't seen anger! But you go ahead and dare him!" (Breault and King, page 123.) Malignant Pied Pipers reveal their apocalyptic potentials when they get wound up during their preaching.

Koresh's insatiable narcissism continued to require more and more sexual adoration from nubile and attractive pubescent young women, whom he conned into believing that they were spiritually privileged and special to have sex with him. The illegal firearms and the increasing evidence of Koresh's sexual abuse of children and adolescents led to the confrontation with the federal authorities that Koresh both longed-for and hated. The overworked Texas Department of Child Protective Services did investigate the Mount Carmel compound. However, due to the cult children's parents' passive narcissism and complicity with Koresh's dominance, the authorities found no obvious evidence of Koresh's deceptive techniques of flagrant child abuse.

One of the key assumptions held by the federal agents during the siege was that the cult members' parental instincts would ultimately override their devotion to Koresh. This assumption proved false. Koresh had so undermined all family relationships that everything revolved around him. In an hour-long interview with one of the mothers who was released in the early days of the siege, she brought up Koresh 24 times, but never once mentioned her husband or their three children, all of whom died in the fire. *(The New York Times,* May 4, 1993.)

To this day the FBI is excoriated for trying to end Koresh's terrible abuse of the cult children. No matter what mistakes the authorities made, at least they had the courage to confront Koresh with the extensive need for appropriate authority, something his sad soul had craved all his life.

As the *Time* special edition concluded, "In the end, even the fiercest critics could not deny that it was Koresh who placed 25 children in harm's way, who preyed on people who were weak and lonely and hungry for certainty. Certainty he gave them, and abundantly. He was certain of his vision of good and evil, certain of his special insight into the deepest mysteries of faith, certain of an after-life that promised glory for those who had suffered for their souls. If he is right about that, and there is any justice in it, Koresh has not seen the last of the flames." *(Time* special edition, page 43.)

Diagnosis of David Koresh

Following the criteria established in the Diagnostic and Statistical Manual (DSM-IV, see page 19 in Chapter One for particulars), as well as the insights offered by Otto Kernberg, I offer the following diagnosis of Koresh.

Koresh meets all nine criteria for a diagnosis of Narcissistic Personality Disorder. His grandiosity, fantasies of power, belief in his own status, insistence on excessive admiration, sense of entitlement, interpersonal exploitation, lack of true empathy, belief that others envied him, and arrogance are well documented. I do not find excessive envy *of* others as part of his personality – like many of the MPPs, he is so grandiose that he would assume everyone envied him or that he was beyond envy.

As to the Kernberg criteria, Koresh meets all four. He became totally regressed, delusional, and paranoid after being literally wounded at Waco. He had shown impulsive and assaultive behavior for many years, weaving his own wounds into his survivalist mentality and potential group assault as part of his daily theology. He had many exploitive sexual relationships and regularly blurred the boundaries of relationships. His hatred of all authority other than his own and his unending narcissistic rage and associated violent behavior relentlessly escalated, culminating in the Waco tragedy.

In addition, he exhibited secondary psychopathic features — statutory rape, assault with a deadly weapon, child abuse, and firearms violations — and possible psychotic symptoms in his final days after suffering bullet wounds in the original gunfight with federal agents.

"His name was death, and hell followed with him."
—Revelation 6:8

CHAPTER FOUR

CHARLES MANSON AND SHOKO ASAHARA: PSYCHOPATHIC PIED PIPERS

"Everybody in the world wants to get mad at me because I won't show remorse because somebody dies. Somebody dies every day."
— Charles Manson (quoted in *The Houston Post,* August 7, 1994)

"It has become clear to me now that my first death will be caused by something like a poison gas such as sarin."
— Shoko Asahara (quoted in *The New York Times,* March 26, 1995)

Why include these two malevolent cult leaders in this book? I include them because they each developed a unique apocalyptic scenario, which they used to build their delusional cult group identity and manipulate their followers. They both used their charismatic personality traits and typical Malignant Pied Piper indoctrination and "mind-control" techniques to influence and lead their followers towards group homicidal behaviors, as opposed to group suicide.

I think both of these men could have been spotted early on — by a medical professional or even by a well-informed lay person — as dangerous because of their bizarre pseudo-theology and charlatanism. It may not have been possible to stop them entirely, but they could have been watched and perhaps challenged before enacting their apocalyptic scenarios. Each of their narcissistic personalities has a decidedly antisocial or criminal psychopathic flavor. There is no better tribute to the memory of the victims of these two psychopaths than for all educated persons to be aware of and recognize the sinister charisma and treacherous techniques of power and control exhibited by typical Malignant Pied Pipers.

Charlie Manson: The "Helter Skelter" Killer

> *"25 years later, Manson says he feels no remorse for murderous rampage."*
> — *The Houston Post,* August 7, 1994

The Critical Incident
During the 1960s, Charles Manson lived in squalor at Spahn Ranch near Los Angeles with his "Family" of teenagers recruited from middle-class California families. Manson mentored his recruits in a life of group sex, petty crime, and scavenging for discarded food. Charlie, like some Fagan-inspired modern Pied Piper, supplied marijuana and LSD for his followers, who thought he was Jesus Christ. Like David Koresh, Charlie would play his guitar; he also told counter-culture stories with a mystic, mesmerizing style, and flattered and primed the young women with drugs and sex. Most of them were already runaways or teens alienated from their middle-class families; all of them were white.

Manson presented his apocalyptic vision of American society's decline into racial war and chaos, using a hodgepodge of Biblical references, Beatles songs, and his own angry anti-establishment lyrics to beguile and entertain his runaway rebel followers, as if he were Christ or a prophet. Manson extolled Hitler, whom he admired and identified with, especially the powerful gaze. Manson, who transfixed his followers with his stare, had a swastika tattooed on the center of his forehead.

At night on August 9, 1969, four members of the Manson Family cult invaded pregnant celebrity Sharon Tate's posh Benedict Canyon mansion while Tate's husband, film director Roman Polanski, was in Europe. These Manson-inspired young people butchered everyone inside, scrawled "PIG" in blood on the front door, and smeared other blood-based graffiti inside. The next night, Manson's minions killed Leno and Rosemary LaBianca, owners of a supermarket chain, in their home in Los Feliz, California.

Manson, Charles "Tex" Watson (age 23), Susan Atkins(21), Patricia Krenwinkel (21), and Leslie Van Houten(20) were convicted of murder and are spending their lives in prison. Manson continues to be an object of morbid media fascination and his face appears on T-shirts and other still-popular memorabilia.

Manson's Apocalyptic Scenario as Malignant "Theology"
Manson repeatedly lectured his white followers about the coming violent revolution of American blacks that would result in their takeover of America's large cities. After their revolutionary victory, Manson said, the blacks would become embroiled in socially chaotic conditions because of their lack of effective leadership. Manson and his elite group would then generously provide leadership and direction to the floundering blacks.

There is evidence that the murders committed by the Manson Family were intended to spawn chaos and spark the black revolution against whites that Manson prophesied. Manson believed that the Los Angeles police would suspect members of the Black Panthers, inciting racial war. This effort at self-fulfilling prophesy is similar to Shoko Asahara's Tokyo subway gas attacks, which Asahara thought would convince Japanese citizens that America and "the West" was invading Japan. (These Pied Piper songs of rampant murder are worse than rat plagues.)

Psychobiography of Charles Manson
Charles Manson was born "No Name Maddox" on November 12,1934, the illegitimate son (like David Koresh) of a 16-year-old

girl named Kathleen Maddox. Unable to remember the date of her son's birthday, Kathleen changed it to November 11, Armistice Day, an easier date for her to recall. Manson has said that his mother was a teenage prostitute. Kathleen's relatives say that she was "loose" and "ran around a lot." She lived with a succession of men, one of whom was a much older man named William Manson, whom she married and hung around with long enough to provide a surname for Charles. Manson stated on a number of occasions that he never met his father. (He, like Koresh and Jones, had his followers call him "Father" or "Dad," and strove to be his strong father/himself.)

According to Manson's mother's relatives, Kathleen would leave Charlie with friendly neighbors for "an hour" and then disappear for days or weeks. Like David Koresh, he spent most of his early years with his aunt or grandmother.

In 1939, Manson's mother and her brother were sentenced to five years in prison for armed robbery. Charles then lived with his aunt and uncle who, like Jones's Mrs. Kennedy, were extremely religious. Manson's aunt thought all pleasures were sinful, but she did give him love. In contrast, Charlie's mother let him do anything he wanted to as long as he did not bother her. Manson's mother Kathleen must have been a source of both shame and excitement for him, because of her status as an armed robber. Manson eventually led a group of young women (who were about the same age his mother had been when she was incarcerated) into a life of crime and imprisonment.

When Kathleen was paroled in 1942, she reclaimed her eight-year-old son Charles from his aunt and uncle. Manson's mother, now 24, soon returned to her alcohol-bathed serial relationships with fellow drunks. Charlie ended up in a series of boys' schools and finally at Father Flanagan's Boys Town in Nebraska because of criminal theft activities.

Four days after his arrival at Boys Town, he ran away with another boy and committed his own first armed robbery at age 13. He ran away from the Nebraska school for boys 18 times.

In 1951, Manson was ordered confined to the National Training School for Boys in Washington, D.C., until he turned 18. His IQ, as tested there, was 109. Although his formal schooling had been haphazard, he had "street smarts" and the capacity to please and manipulate teachers and other adults. A worker at the school observed the following about Manson after three months: "Manson has become somewhat of an 'Institutional Politician.' He does just enough work to get by on … restless and moody, the boy would rather spend his class time entertaining his friends." *(Helter Skelter,* Vincent Bugliosi and Curt Gentry, page 187.)

In 1951, at age 17, Manson was examined by a psychiatrist. Dr. Block talked about the significant degree of rejection, instability, and psychological trauma in Manson's background. Dr. Block noted that an inferiority feeling caused by his relationship with his mother was profound. (I see his mother as a source of shame, humiliation, and narcissistic injury requiring over-compensation via grandiosity and haughtiness.) Block observed that Charlie found it necessary to totally suppress or deny any thoughts about his childhood. (Massively denied shame, humiliation, and rage can reemerge later as violent actions.) Psychiatrist Block felt that Manson's short stature, his illegitimacy, and the lack of parental love in his childhood led him to constantly seek approval or status from his peers. Block wrote: "To attain this (status), Manson has developed certain facile techniques for dealing with people. These for the most part consist of a good sense of humor and an ability to ingratiate himself … This could add up to a fairly institutionalized youth, but one is left with the feeling that behind all this lies an extremely sensitive boy who has not yet given up in terms of securing some kind of love and affection from the world."

Kindly Dr. Block was conned. The way Manson would get "affection from the world" was through the circus-like, flash-bulb-popping glow of the national media and publicity for his delusional, narcissistic speeches. Manson sounds psychotic, but I think he is crazy like a malignantly narcissistic fox.

Another Psychiatrist Gets Conned

In 1955, when he was 21, Manson stole a car in Ohio and drove with his pregnant wife to Los Angeles, where he was arrested in October of that year. Manson claimed that he was badly in need of psychiatric treatment and had stolen the car as a means of mental catharsis for his confused state of mind. On October 26, 1955, the judge ordered a psychiatric evaluation of Manson. Dr. Edwin McNeal diagnosed an "unstable personality"; although finding Manson a poor risk for probation, this psychiatrist optimistically recommended it anyway. McNeal felt that a wife and impending fatherhood might provide motivation for Manson to "straighten his life out." Charlie skipped out on his probation and was back in jail by the time Charles Manson Jr. was born. So just as his mother had done to him, Charlie spent his own child's childhood in jail or committing progressively more serious crimes. (*Helter Skelter,* pages 190-191.)

Manson's ability to deceive the concerned Dr. Block at the National Training School for Boys and later his "Helter Skelter" Manson Family murder cult are part of his gruesome history. We can learn a lot from Manson about how to avoid the clutches of a Malignant Pied Piper. Teenagers and their parents go through a very vulnerable period as the young people push away from their parents as part of a normally rebellious search for independence. Unusual, "weird" counterculture heroes that look and act strangely charismatic can appeal to teenagers' need to be "different," which they equate with being independent.

Manson's domination and acted-out degradation of his women cult members is typical of all Malignant Pied Pipers. Manson used his charismatic qualities to psychologically engineer the rebellious participation of these young women in the mutilation and murder of a beautiful pregnant woman. This crime reflected Manson's long-sequestered shame, humiliation, and rage at his mother. Prison, which had been his mother's home when she should have been creating a decent home for Charlie, has now become his permanent home, by his own admission.

Manson's evil soul has basked in the spurious, illusionary restitution/revenge toward his mother that was displaced onto Sharon Tate. The murder of the Tate mother and baby insured the reality of his cult daughter/women's incarceration. Manson's cult women, like his mother, would spend endless, boring hours, days, and years in prison. Manson illustrates the perverted attempts at restitution and revenge that characterize the unconscious acting-out of destructive cult leaders.

Material from Manson's comments at media interviews confirms the notion that he victimized the middle-class runaway girls of an American society swept up with materialism and selfish greed. (Osama bin Laden would give high-fives to Manson's concept of a sick America!) (See *Helter Skelter,* pages 653-656.)

We find similar derisive comments in Jim Jones's put-downs of his followers in his "white night" sermons in Jonestown in the Guyana jungle. Malignant Pied Pipers illustrate what I have called "the malignant transformation of narcissism" in cult leaders and the passive malevolent surrender of their individuality and personal authority in cult followers. The cult leader/preacher puts down his followers as he simultaneously builds up their special status, which will be attained by carrying out his exciting projects.

Rather than experiencing the usual altruistic surrender to a worthy and benevolent cause, as described by Anna Freud in normal, idealistic, and religious youths (Freud, Anna, *The Ego and the Mechanisms of Defense*, page 123), Manson, Koresh, Asahara, Jones, and Jouret surrendered malevolently to the stored-up inner rage at their parents and a society that they felt had keenly disappointed them. They struck back in cowardly fashion at society, like the bitter, "stiffed," vengeful Pied Piper of Hamelin, who took revenge on the village by seducing away the village's children with the music of his cult.

Diagnosis

Charles Manson fills all DSM-IV criteria for antisocial ("criminal psychopath") personality disorder. He sounds psychotic in his self-serving, self-luxuriating ramblings, but I see these as evidence of Narcissistic Personality Disorder and the ultimate in Kernberg's notion of Malignant Narcissism. "Extreme evil" also fits.

Manson claimed to be God or Christ but this was clearly his own brand of blasphemous and mocking iconoclasm, a point of distinction from Jones and Koresh, for example. He recruited approximately 25 people to his cult, but it is difficult to be precise because some drifted off or remained missing or unaccounted for. Manson himself bragged about committing 35 murders, but many could never be proven.

A "Group Self Diagnosis" of the Manson "Family"

"... but at this time we might suggest the possibility that she may be suffering from a condition of 'folie a family,' a kind of shared madness within a group situation."
— From Dr. Joel Hochman's psychiatric report on Susan Atkins *(Helter Skelter, page 623)*

I believe psychiatrist Hockman is on target with his observations about the Manson family. Hochman described a "double process of selection" for membership in the Manson family. The "in-betweener" neediness in the followers reverberated with Charlie's needs to be a powerful, Christ-like father and omnipotent leader. Bugliosi describes Manson's use of drugs, repetition, isolation, fear (those penetrating eyes that Charlie bragged were like Hitler's), sex, and love. The love emerged out their sharing of communal problems and pleasures.

"They were a real family in almost every sense of that word, a sociological unit complete to brothers, sisters, substitute mothers, linked by the domination of an all-knowing, all-powerful patriarch. Cooking, washing dishes, cleaning, sewing—all the chores they had hated at home they now did willingly because they pleased Charlie." *(Helter Skelter, pages 654-655.)*

Shoko Asahara: Malignant Pied Piper of Japan

"Don't touch me, I don't even let my followers touch me."
— Asahara to police doctors after his arrest (quoted in
the *Keene [N.H.] Sentinel*, May 16, 1995)

Critical Incident and Background

During the morning rush hour on March 20, 1995, Asahara's Aum Shinrikyo cult members released deadly sarin gas in the Kasumigaseki station of the Tokyo subway system. Twelve people died and 5,000 were sickened by this cult group action. At the time of the attack, the cult had 10,000 followers in Japan and as many as 40,000 members in Russia. Before the attack, Asahara appeared regularly on Japanese television, and in 1990, he and other cult members ran (mostly without success) for seats in the Japanese Parliament.

Asahara's Psychobiography

Shoko Asahara was born in 1955 on Kyushu, one of Japan's main islands. Born sixth of seven children, Asahara was sightless in one eye from birth and partially blind in the other eye. His father was a craftsman whose trade was making straw mats. Asahara's visual problem prevented him from learning his father's trade. His parents decided to send him to the distant city of Kumamoto where he could attend a state-subsidized school for the blind. His younger brother, whose eyesight was normal, also gained admission to the same school. Asahara never returned home again.

At the school for the blind, Asahara discovered a peculiar power because of his limited ability to see. A former teacher said, "Being able to see even a little is prestigious because blind children want to go out and have coffee or tea at a tea room, but can't go by themselves. They would say to Chizuo (Asahara), 'I will buy you dinner, why don't you take me out?'" (Van Biema, David, "Prophet of Poison," *Time*, April 3, 1995, page 30.)

This position of social power allowed him to be a leader; he sought power by becoming the champion of social outcasts at the school. At times he also was described as a bully. An ex-teacher remembered an occasion when Shoko Asahara was reprimanded; in retaliation, he threatened to burn down the school dormitory. When confronted, he backtracked, saying that he couldn't be punished for just saying words. Asahara inflicted a broken eardrum on one classmate in a fight. Later, in high school, he lost every attempt at class office because classmates were frightened of him. These ballot-box defeats were at odds with his frequent claims that he would become prime minister of Japan someday. (Van Biema, page 30.)

Asahara was rejected for study at Tokyo University. (This was the first in a series of dark epiphanies.) In true Malignant Pied Piper fashion, Asahara lured many bright young students from Tokyo University into his murder cult. (Just as Charles Manson lured young woman into his cult, which eventually led them to crime and life imprisonment, Asahara's acted-out hatred of Tokyo University became complete when his student followers ended up in jail awaiting execution.)

When Asahara was denied admittance to Tokyo University, he became an acupuncturist. But he was arrested in 1982 for selling fake cures. (I see this as another Dark Epiphany.) He was detained in jail for 20 days, was fined 20,000 yen, and his business went bankrupt. Rejected by the university and rebuked by the legal system, he turned to the next best thing, telling a woman assistant, "I believe the future lies in Religion." (Sayle, M., "Letter From Tokyo: Nerve Gas and the Four Noble Truths," *The New Yorker,* April 1, 1996, page 58.)

In 1984, at age 29, Asahara went to India to study Buddhist enlightenment. Six years later, Asahara returned to Japan to found "The New Society of Aum." Aum is the same as "Om," a Sanskrit mantra representing the three major Hindu gods. He recruited members by making speeches and appearances at Tokyo University and Buddhist temples.

As is so typical of our Malignant Pied Pipers, Asahara told his followers that first Japan and then the whole world would acclaim him as the new Buddha and the savior of humanity. One of the four noble Buddhist truths involves chastity, but like Jones, Koresh, and Applewhite, only the followers were to be chaste. Asahara himself lived comfortably with his wife — an adoring follower. (She is now in prison and relatives are raising their six children.)

Even as a child, Asahara had grandiose fantasies of becoming a political leader. When he ran for office in school he was rejected; Japanese voters rejected him (though some of his followers got elected to parliament); Tokyo University rejected him. These accumulating Dark Epiphanies led to escalations in Asahara's apocalyptic predictions of Armageddon. When the United States did not cooperate and attack, thus propelling Asahara and his cult into roles as the great protectors of Japan, Asahara made his own prophesy come true in a ghastly way by his gas attack. This strongly resembles bin Laden's attack on Saudi Arabia when the Saudis rebuffed him. It is the way Manson's "Helter Skelter" vision turned cult members into murderers to make his narcissistic power-grab come true. Applewhite, Jouret/DiMambro, Jones, and Koresh all utilized various forms of these self-fulfilling prophecies to promote their escalating delusions.

Asahara's Young-Adult Pied Piper Song

Asahara benefited from the boom of new religions in Japan in the 1980s. His self-expression turned messianic in 1987 when he declared himself to be "Today's Christ" and "The Savior of This Century." His "new" religion was nominally Buddhist, but was really a hodgepodge of ascetic disciplines, Christian platitudes, and New-Age occultism. Asahara began to identify supposed threats to Japan from the United States, Jews, and Freemasons. His ideology and bizarre theology, like that of Koresh, Jones, and Manson, grew steadily toward apocalyptic visions of impending doom.

Early in 1995, Asahara published a book called *Disaster Approaches the Land of the Rising Sun.* He likened himself to Hitler in his passionate nationalism, and once spoke to 15,000 people at a Russian sports stadium. In his sermons, Asahara constantly wove his apocalyptic doomsday scenario into the actions and expectations he predicted for his followers. This of course is typical of Malignant Pied Pipers.

Asahara hit the Japanese paranormal media big-time when the Japanese magazine *Twilight Zone* ran a photo of him meditating in the lotus position while apparently floating in mid air. *(The New Yorker,* April 1, 1996, page 59.) Fortuitously for him, Asahara and his cult came to prominence in the middle of a Japanese generation gap crisis. Sayle aptly notes that Asahara's parents' generation had never experienced anything but drudgery, war, defeat, and hunger. By the time Asahara came of age in the 1970s and 1980s, many Japanese young adults were beginning to see the custom of working faithfully in one company for life and living with "out-of-touch" parents until marriage as a real "death-in-life." Asahara's cult offered these restless young adults not only jobs, but also a common cause and exciting communal living.

Japanese TV of the time drew huge audiences to science-fiction serials depicting Godzilla lizards spawned by evil powers from outer space. The uncanny parallel and resonance between Asahara's "shtick" and this pop culture was that the "young warriors of truths" in these sci-fi movies fought against the "external" powers that sought to turn the common people into robots. The warriors of truth in these movies started to hum in tune with Asahara's Pied Piper rebellious music. Thousands of bright Japanese youths began to flock to this exciting new, modern Japanese father/older brother anti-hero named Asahara.

Susuma Oda, a professor of psychopathology at the University of Tsukuba, feels that one of the attractions of cults for Japanese youth is analogous to the lure of drug culture for Americans. These young people's fathers were always at the office and seldom home with their family. His theory is shared by several other Japanese sociologists and observers. *(The New York Times,* March 26, 1995, page 8.)

Asahara and his lieutenants used the usual tools of Malignant Pied Pipers. These included psychological manipulation, social coercion (banning of sex except with him), and limits on reading and television watching except for Asahara's writings, tapes, and sermons. He used sleep deprivation and hallucinogenic drugs, and he took over cult members' financial assets. (Van Biema, *Time*, April 3, 1995, page 31 and Kristof, Nicholas, *The New York Times*, March 26, 1995, page 8.)

Asahara's Narcissism Confronted

In October 1989, *Sunday Mainichi*, one of Japan's biggest selling magazines, began a series of articles on Aum Shinrikyo and Asahara entitled "Give Back My Child." This article featured six families who charged that Asahara had stolen their teenagers. Asahara also began escalating his coercion of recruits, using methods that included kidnapping, murder, and eventually the gas attacks. Finally the Japanese authorities took effective action and arrested Asahara.

The legal process and investigations took a long time because Asahara had cleverly developed degrees of distance from the actions of his followers, in a way that is reminiscent of Manson. On February 27, 2004, Asahara was finally found guilty of 27 murders after an eight-year trial. He will make many appeals. It took the judge four hours to pronounce guilty verdicts on thirteen charges as Asahara crossed his arms, smiled, yawned, snorted, scratched his head, smelled his fingers, mumbled, and recited mantras. *(The New York Times*, Feb. 27, 2004.)

Asahara's pathologic narcissism was reflected in the fact that he always sat one level higher than his followers and they had to kiss his toe when they had an audience with him. A terse summary of Asahara's maniacal conative core and fantasies about his destiny comes from one of his schoolmates from his early school years: "I think that Asahara is trying to create a closed society, like the school for the blind we went to. He is trying to create a society in which he can become king of the castle." (Van Biema, *Time*, page 31.)

Diagnosis: Asahara's Personality Disorder

In my opinion Shoko Asahara has a "Mixed Personality Disorder" without meeting full criteria for Kernberg's notion of Malignant Narcissism (he fits only # 3).

Narcissistic Personality Disorder (Asahara shows 8 of 9 of the classic criteria as outlined in the DSM-IV).

1. Grandiosity: In 1987 Asahara declared himself to be "Today's Christ" and "The Savior of this Century."

2. Preoccupation with fantasy: In 1995 Asahara published a book, *Disaster Approaches the Land of the Rising Sun*. His religion (an amalgam of Buddhist, New Age, Hindu, and ascetic teachings) would provide refuge from his predicted attacks by America, Freemasons, and Jews. (*The New York Times*, March 26, 1995, page 8.)

3. Special Status: Asahara likens himself to Hitler (as well as Christ and Buddha) and once spoke with some kind of inflated identity to 15,000 people at a Russian sports stadium.

4. Need for excessive admiration: In Ian Parker's description, "Live-action TV footage (at his performances) depicted the guru (Asahara) apparently performing miracles like levitation, while animated films showed him flying through cities and passing through walls. Other footage showed Aum rites: young adherents of both sexes — clad in white satin and wearing, with Japanese sensitivity to status, colored sashes indicating religious rank — approaching the tubby guru, resplendent in see-through golden silk robes, prostrating themselves, and reverently kissing his toe." (*The New Yorker*, April 1, 1996, page 60.)

5. Entitlement: This feature seems implied in Asahara's approach to recruits, the media, and even the Japanese government until his sarin gas attacks were exposed.

6. Exploitive: In 1989 Aum was recognized as a religious body in Japan. It claimed 4,000 members (380 "ordained"). In 1995 a U.S. Senate committee estimated the worldwide following as 50,000 and financial assets at

more than a billion dollars. Thirty thousand people had joined the Aum cult in Russia. "Ordination" in Aum was reserved for landed or scientifically talented elite. "This required taking vows of chastity; cutting all ties with the world; renouncing families; and signing over worldly possessions to the cult, including real estate, savings, clothes, telephone calling cards, and personal seals." (Typical MPP "bottom lines." Asahara's exploitation of disaffected upper-middle-class youth wanting to get away from humdrum parents is very similar to that of Manson.) (Sayle, *The New Yorker*, page 60-61.)

7. Lacks empathy: As usual with MPPs, any perceptiveness or empathy in Asahara served his own purposes.

8. Envy: As usual—the whole world would envy his wonderfulness.

9. Arrogance: Very obvious.

Kernberg's Malignant Narcissism:

#3. Psychopathy: Like Jones, Koresh, and Manson, Asahara has plenty of antisocial, criminal, and psychopathic traits. In 1982 Asahara was arrested for selling fake cures. He paid 200,000 yen ($800) and went bankrupt. This spurred him on to Elmergantryville enterprises, including murder. *(Time,* page 31.)

Next stop, Heaven's Gate and the Solar Temple.

CHAPTER FIVE

NOT OF THIS EARTH:
MARSHALL APPLEWHITE OF HEAVEN'S GATE;
LUC JOURET & JOSEPH DIMAMBRO
OF THE SUICIDAL SOLAR TEMPLE

Marshall Applewhite: An Extraterrestrial Pied Piper at Heaven's Gate

> *"Planet Earth About to Be Recycled. Your Only Chance to Survive — Leave with Us"*

> *"The true meaning of 'suicide' is to turn against the next level when it is being offered."*

> —"Do" (Marshall Applewhite), leader of Heaven's Gate *(Time*, Special Report, "Inside the Web of Death," April 7, 1997)

The Critical Incident
How could 39 intelligent, idealistic, and dedicated people quietly allow themselves to die under the dominion of a guru who claimed that their spirits would be transported to a spacecraft traveling behind the Hale-Bopp comet? It may seem bizarre, yet these well-

educated people did just that. The members of the "Heaven's Gate" cult so believed in their Pied Piper's "Music Man" song that they bought into the incredible idea that he would lead them to higher, spiritually rich domains of existence in outer space. On March 26, 1997, Marshall Applewhite, age 66, and 39 followers died together in a mansion at Rancho Santa Fe, California.

Applewhite had been theologizing, proselytizing, and preaching a blend of Christian astrology and outer space fantasy since the 1970s. Applewhite promised his disciples that they would evolve into spiritually superior, bald aliens by severing all links to modern society and human desires. Heaven's Gate suicide cult members not only shunned sex, but approximately a third of the men in the cult chose castration, following Applewhite's personal example.

The 39 followers died by drinking a lethal ingestion of phenobarbital and alcohol, while being suffocated by plastic trashbags tied tightly over their heads. They left goodbye messages on videotapes. All wore black clothes, new Nike shoes, and had identification cards pinned on their clothing. They all wore triangular purple shoulder patches with the name Heaven's Gate sewn on them. And they all apparently believed Applewhite when he told them their human bodies were temporary vessels and their deaths were timed to rendezvous with an unidentified flying object trailing the Hale-Bopp comet.

Applewhite's psychotically depressed hunger for personal power, control, and affirmation dominated his later days. With Bonnie Nettles, he studied and produced a strategy of charismatically presented ideals directed at eliciting sexless devotion from followers. Applewhite and Nettles created theologically based propaganda about a literal extraterrestrial flight towards the ultimate denial of sex and death.

Applewhite's Psychobiography
Applewhite wasn't always the strange, sexless alien guru the world would meet when they saw his picture on the cover of *Time* magazine. According to his younger sister, Louise Winant,

Applewhite had been a totally normal man until he had a life-changing, near-death experience in a Houston hospital.

Marshall Herff Applewhite was born in 1931. He was the son of a Presbyterian minister who specialized in starting new churches, so his family moved from place to place in Texas. As a teenager, Applewhite wanted to become a preacher like his father, but decided to pursue his talent in music.

Throughout his 30s, in the 1960s, Applewhite lived a very conventional (if repressed) life. He married and had two children. His sister Louise described him as a "very loving and wonderful brother." Early in his life, Applewhite seemed to have been struggling with neurotic conflicts and inhibitions that can occur with a "plain vanilla" family situation. His unconscious homosexual conflicts were apparently held in check by repression until midlife, when they started coming out.

Applewhite sang in starring roles in local stage musicals in Houston and Colorado. He taught music at the University of St. Thomas in Houston and was choir director at St. Mark's Episcopal Church in Houston. (*Time,* April 7, 1997, pages 40-41.)

Beneath the "normal" surface described by his sister, dark epiphanies were lurking and accumulating. In 1970, he got a big break. He got the role of lead baritone in the American opera "The Ballad of Baby Doe" in New York City. The part was too much for Applewhite and he didn't have enough voice for the difficult role. According to Charles Rosekrans, the choirmaster, Applewhite handed him a letter from a psychiatrist before withdrawing from the production. (*Time,* page 40.)

Back in 1964, Applewhite had been fired from a music position because of homosexual approaches he made to male students. Then in 1971 he was dismissed from his teaching job at St. Thomas because of a homosexual affair with a student. He checked into a psychiatric hospital "to be cured of his homosexuality." The repressed homosexual conflicts were clearly surfacing when he had the added anxiety of medical problems. He had a heart attack and spent some time in the intensive care unit, but recovered. He was hospitalized briefly for psychiatric reasons, probably suffering from post-coronary depression.

After his release, Applewhite confided to a lover that he longed for sexless devotion and passion without physical entanglements. He left his wife and two children. He began to evolve his own narcissistically derived healing myths. He began to talk about influences from "the next level" and told others he felt he was "one of *the two*." (He clearly had begun the typical process of Malignant Pied Pipers — i.e., projecting his own inner conflicts and unresolved sexual conflicts onto his followers via the evolving theology and content of his cult doctrines and apocalyptic pronouncements.)

The other half of the "two" was Bonnie Lu Nettles, then 44, a nurse who had attended Applewhite in the hospital. Nettles told Louise Winant that Applewhite "had a purpose when God had kept him alive."

"Their relationship wasn't like a romantic thing, more like a friendship, a platonic thing," said Bonnie's daughter, Terrie, when interviewed by CNN Impact's Henry Schuster and *Time's* Patrick Cole. Nettles, who embraced astrology, believed that she and Applewhite were fated in the stars, according to Terrie Nettles.

"A couple of spiritualists had said that there was going to be this guy coming into her life, but he had not shown up in the past for Nettles. Now, enter Marshall Applewhite. They linked together on a spiritual plane," explained Terrie Nettles. No specific information concerning Bonnie Nettles' sexual orientation or sexual contentment is available, but one would speculate that there was resonance on her part with Applewhite's profound disappointment in and shame with heterosexuality.

In any event, they merged narcissistically on a "spiritual plane." According to the *Time* account, Applewhite saw the relationship with Bonnie Nettles as some special spiritual destiny. Bonnie Nettles had written a novel in junior high school about a man who died and went to heaven. Applewhite grabbed onto Nettle's literary vision as a prophecy. He claimed spiritual ownership of the novel's hero as his ascendant new identity.

Before they reconnected in the hospital's coronary unit, Bonnie Nettles had been Applewhite's music and drama student in Houston. She drew up his astrological charts and "channeled her spirit advisor, Brother Francis" for guidance. In 1972, the Applewhite-Nettles partnership culminated in their starting a New-Age school called the Christian Arts Center, which taught astrology and metaphysics. It was here that Applewhite's intensity and charm intensified into full-blown spiritual charisma. And it was well timed. In the late 1960s and early '70s, Houston saw a resurgence of New Age self-help groups that used "spirituality" as their message. After a failure in music and an acute medical scare, Applewhite turned back to being a preacher, a vocation he had once rejected.

"I felt like I was in the presence of an incredible human being. It was as if I was being uplifted," commented Terrie Nettles. "I felt privileged to be with my mother and him. I was the only one who could talk to them together. Their followers had to talk to them in groups, not individually." Here is a key warning sign of a Malignant Pied Piper! Any healthy religious leader can relate clearly and comfortably to individuals or groups without special circumstances, controls, or special protection for his or her ideas.

By 1973, Nettles and Applewhite were convinced in their resonating narcissistic twinship that they were the Two Witnesses prophesied in the Biblical book of Revelation, Chapter 11. They fervently believed that it was their destiny to prepare *the way* for the kingdom of heaven.

They began to travel around the country together. In her grandiose style, Bonnie Nettles wrote to her daughter Terrie: "I'm not saying we are Jesus. It is nothing as beautiful as that, but it is almost as big. We have found out, baby, that we had this mission before coming into this life. All I will say is it's all in the Bible, in Revelation." (The book of Revelation was also used extensively and articulately by David Koresh to expound his prophesies and apocalyptic predictions.)

In 1975, Bonnie Nettles and Applewhite left Houston permanently. Mother and daughter never saw each other again. Applewhite abruptly severed all ties with his own family. His sister,

Louise, confirms the absolute cruelty of the departing narcissist when she said simply but profoundly, "He hurt his family and children very deeply." Applewhite spent no time explaining his vision to his wife or children. They were apparently not even allowed to hear about his special new calling. After the Heaven's Gate apocalypse, Applewhite's son (a minister) apologized to the world for his father's tragic deeds.

It seems likely that Applewhite's narcissism had led to genuine neglect of his children's psychological needs long before his mid-life crisis. Retrospectively, the tsunamic proportions of his own pathologically narcissistic needs, insecurities, and conflicts had been building for years. The way he coldly cut his ties with his family is a pivotal event and part of his build-up of dark epiphanies. Other contributors include his repeated trouble with improper familiarity toward male students; rejections for better roles at the Houston Grand Opera and being panned by local music critics; his coronary and subsequent depression amidst a mid-life crisis; and his repression of his homosexual "coming out."

Applewhite's abandonment of his own wife and children was a harbinger of worse abuse, abandonment, and cruelty to his followers in the future. At least his own children were in that sense saved from the ultimate bogus con-job promoting suicide as mystic entry to paradise. In this sense, Applewhite's suicide con was similar to but on a lesser scale than Osama bin Laden's bogus promises to his cult members about Allah's special favors after martyrdom or death in Jihad.

Here lies the core pattern in destructive cult leaders: They seduce their followers with the utopian promise of empathetic resonance and healing empathy, but they end up hurting, neglecting, and killing the hungry followers, just as they themselves had felt abused, neglected, or hurt in their childhoods. Rather than accept or completely deal with his conflicted homosexual desires and compulsions by therapy, Applewhite used denial and acting out of his inner pain through otherworldly escapism. He could have joined the openly gay community in Houston's Museum/Montrose area.

Many therapists were helping gay persons to "come out" during the 1970s and '80s. He could have dealt with his family's pain over this and found an ordinary and productive life. Instead he began to search for the stars via denial and grandiose delusions.

Bonnie Nettles and Applewhite traveled around the country and eventually settled in Los Angeles, where they first called their group Guinea Pig, with Bonnie Nettles designated as "Guinea" and Applewhite as "Pig." However, soon the group came to be called the Human Individual Metamorphosis, and Applewhite was designated "Bo" and Bonnie Nettles "Peep" in reference to their roles as cult shepherds. Later, they were called "him" and "her," then finally, the musical "do" and "ti," the first (Applewhite) and seventh (Nettles) notes of the musical scale.

The early days of the Bo-Peep cult were unlike the later high-tech Rancho Sante Fe version of Heaven's Gate. Applewhite and Bonnie Nettles did odd jobs to support themselves and were even arrested in Harlingen, Texas, for stealing gasoline credit cards. Applewhite stole a car and spent months entangled in legal conflicts about the event.

During the adolescent phase of the cult, Applewhite wrote his first spiritual manifesto. (Most exploitive cult leaders either write their own theology or reinterpret some traditional scripture to suit their own grandiose purposes of power and control.) During this time Bo and Peep, while stranded in a broken-down car, prayed that God would provide a rescue for their predicament. On that same night, a car picked them up and the comet Kohoutek appeared in the clear night sky. The much-hyped comet fit into their expectations and prophecies of UFO's coming to earth, another extraterrestrial solution to their inner problems.

Bonnie Nettles and Applewhite continued to preach passionately and persuasively to many converts, insisting they should renounce their families, sex, and drugs with a higher goal of pooling resources for a future voyage of salvation on a space ship. The gurus also asked their followers to do what they themselves had done, such as renounce their own families and deny the importance of sex. In

Applewhite's case, he had himself castrated and asked others to follow suit. This ultimate destructive projection by the cult leader on to his "new family" in the cult is typical. In this case, Applewhite's promiscuous homosexual intrusiveness and unconscious shame were the source of his projected need to advocate castration as the radical, albeit pathetic, means of sexual ethics and control. Many of his male followers — both gay and straight, but apparently conflicted about sexual and gender issues — followed his example and were castrated.

"I'm familiar with irreversible steps...some students had chosen...to have their vehicles neutered...I'm one of those...and I can't tell you how free that has made me feel." — a male follower of Applewhite *(Time,* page 29.)

Nettles and Applewhite's media sophistication grew. A poster for an appearance in Redwood City, California, read, "If you have ever entertained the idea that there may be a real, physical level beyond Earth's confines, you will want to attend this meeting." The auditorium was packed that night.

The cult came to national attention in 1975 after two dozen people from the small town of Waldport, Oregon, dropped everything to follow Bo and Peep, traveling first to Grand Junction, Colorado, where Applewhite expected to rendezvous with a UFO. Many of those who joined the cult were already seekers of some sort. "We were young longhairs, escaping the madness of Vietnam, the craziness of the 1960s," explained Aaron Greenberg of Waldport, who had moved there after a stint in Vietnam. "The timing was just right. He had knowledge. He had power. Separation of body and soul was a key teaching." A 1975 *Time* article described Applewhite as having a "rare ability to impress audiences with the urgency and truth of his message." Bo and Peep's appeal was so great at the time that NBC aired a pilot series about an extraterrestrial couple called "The Mysterious Two," originally entitled "Follow Me If You Dare."

Bo and Peep's disciples, however, were not all sheep. Many discontented followers rejected the cult when the promised mountaintop flying-saucer spaceship never materialized in

Colorado. After such a healthy challenge, Bo and Peep started a "boot camp" phase of their development allegedly to prepare followers for the rigors of space travel. Family contacts at the boot camp were frowned upon — another typical cult tactic designed to isolate cult members. (See Chapter Six for more cult recruitment tactics.)

In 1984, Bonnie Nettles contacted her daughter after a long silence and said she was so deep into the cult movement that she didn't know how to get out. In 1985, she sent her daughter $200, indicating she would soon be leaving on a UFO. Instead, Bonnie Nettles died of cancer later that year. Her last message to her daughter seemed to reflect both her profound denial of physical death and also her preconscious awareness of her own impending demise. Months went by before her family found out she was dead.

After her death, Applewhite always left an empty chair for his deceased partner-in-exploitation during the in-group meetings. He referred to Peep as if she were hovering nearby. It was soon after Bonnie Nettles' death that Applewhite was castrated, along with five other male followers. He never contacted or empathized with Bonnie Nettles' family.

In the 1990s, the group renewed its recruitment drive, using a more sophisticated appeal and attracting a crowd of mostly well-off and well-educated entrepreneurs. The Heaven's Gate cult members earned money by designing commercial Web sites (including an elaborate one for their own cult) and doing other work in the high-tech industry. They adopted identical costumes and androgynous haircuts that made it difficult to tell the males from the females.

The lives and deaths of Heaven's Gate cult members are unbelievable, but also important to understand. Human beings all strive for meaning and significance in their lives. A charismatic and confident leader appeals to many. The less the followers possess charisma, the more attractive they find it in their leaders. All are particularly vulnerable to cult recruitment during what Singer has called the "in-between" phases of life, such as between high school and college, between jobs, between marriages, between positive relationships. Many of the suicide notes left by Heaven's Gate cult members illustrated their susceptibility and disaffection from normal life.

Most exploitive cult leaders have been so disappointed in the lack of their parents' leadership that they seek to be leaders to other people. Inevitably, these fragile narcissists end up recapitulating the same painful neglect, abandonment, abuse, and loneliness with their followers that they had originally experienced. In Applewhite's case, he does not seem to fit this mold. He came out of a classically repressed, strict religious family. He differed from other Malignant Pied Pipers in that his reluctance to deal with his inner neurotic homosexual conflicts led to his acting-out with his cult followers as pawns in his elaborate fantasies and defenses.

Escalating Dark Epiphanies
Applewhite had severe neurotic conflicts about his own troubled sexuality and gender identity. His sexual acting-out lost him his job as a professor of music. He had some success as a performer at the Houston Grand Opera and the Gilbert and Sullivan Society of Houston, but experienced rejection by local music critics and failure in the "big time" of New York City. By personal communication with his opera colleagues and several personal social experiences with Applewhite, I formed a picture of an extremely flamboyant, entertaining, but self-absorbed man with less talent than he regarded himself as possessing. He seldom got good reviews in the Houston papers and his wife rarely attended his performances. His career frustrations and disappointments very likely left him feeling empty, and hungry for affirmation. These "slings and arrows of outrageous fortune" combined with his probable lack of marital intimacy to propel a typical midlife crisis into a midlife depression and psychiatric hospitalization after a heart attack.

He projected a total renouncement of sexuality into the delusional theology of his New Age religious cult and asexual merger with Bonnie Nettles. He attempted to reverse his personal sexual failures by preaching the abnegation of sexuality to his followers, as if it were some superior spiritual plateau paralleled only by the ancient Gnostics. When Bonnie Nettles died, Applewhite's failed salvation effort through their spurious healing merger imploded. He escalated his extraterrestrial, delusional denial of death into his own soft-sell version of a group suicide as a special escape to better worlds.

Apocalyptic Scenario
Applewhite's apocalyptic scenario, which became woven into his daily sermons to the Heaven's Gate cult, was that their special group awaited a transcendent step beyond Earth. They would, he promised, arrive at a spot in the universe where a higher plane of spiritual existence awaited them. Applewhite convinced the cult members (with their complicit masochistic narcissism), that they all must commit suicide together so that their spiritual selves would join a spacecraft that would fly them away in the tail of the Hale-Bopp comet to their new spiritual destination. These cult members were above average in IQ and were successful people in many respects before they joined Applewhite's special cult. In this respect they were idealistic searchers, similar to DiMambro and Jouret's Solar Temple recruits.

Unlike most of the other Malignant Pied Pipers described in this book, Applewhite apparently did not claim sexual partners for himself from among his cult followers. His severe conflict over his own sexuality led to a delusional level of behavior that involved such asceticism and denial of sexuality that he had himself castrated and convinced many of his male followers to do that same. This denial and projection probably also played a part in his bizarre, otherworldly theology. He put this seemingly idealistic saleman's spin on a sinister apocalypse. It is fortunate that he had not recruited any more members than he did.

The foul-smelling bodies at Rancho Sante Fe spoke countless words about Applewhite's unique but poignant denial of depression, sexual confusion, rage, and death.

Diagnosis

Applewhite fits all of the criteria for Narcissistic Personality Disorder: Grandiosity (his messianic identity as a spiritual leader), persistent fantasies (non-bodily space travel, belief in UFOs), special status, need for admiration, entitlement, exploitive, lacks empathy (abandonment of his family), object of envy (like bin Laden, he grandiosely seemed to feel himself to be above envy), and arrogance.

In the DSM-III (now reworked and revised as DSM-IV) he would have met full criteria for "Ego-Dystonic Homosexuality." In the first edition of his excellent and readable textbook *Essential Psychopathology,* Jerrold S. Maxmen says several important things:

"In defining homosexuality, the clinician should distinguish between sexual behavior, fantasies, and arousal. Homosexuals commonly experiment with heterosexual relations; some marry 'to become straight', to deny their homosexuality, to please parents/ society, or to have companionship, and especially, to raise children. Crime rates, including child molestation, are no higher among homosexuals." (page 273)

Maxmen goes on: "Homosexuality is only considered a mental disorder when the person persistently and intensely dislikes his own homosexuality. DSM-III calls this 'ego-dystonic homosexuality' (EDH), and defines its essential features as: a) heterosexual arousal that is persistently weak or absent, b) homosexual arousal that significantly interferes with heterosexual relations, and c) homosexual arousal that is a chronic, unwanted source of inner distress." (page 273)

In DSM-IV, all concepts of ego-dystonic homosexuality were dropped, I think out of political correctness. In my opinion, Applewhite's diagnosis would have to include ego-dystonic homosexuality to such a degree that it led to a midlife crisis, which merged into a chronic escalating psychotic condition (major depression with psychosis). He had some antisocial behaviors but they did not fulfill criteria for antisocial or psychopathic personality disorder. In his final months, I suspect he would have met criteria for major depression with psychosis — with repressed grief for Ti.

Joseph DiMambro and Luc Jouret's Bad Medicine: Pied Pipers of the Suicidal Solar Temple

Plato and Socrates say, "What we wish to recognize is the following: surely some terrible, savage and lawless form of desire is in every man, even in some of us who seem to be ever so measured. And surely this becomes plain in dreams. Of the tyrant, the truly evil man, he says ... what he had been in dreams, he became continually while awake. He will stick at no terrible murder, or food or deed. The worst man is, while awake, presumably, what we describe a dreaming man to be.
— Marty Stein, "Dreams, Conscience and Memory" (*Psychoanalytic Quarterly,* 1991, LX, page 199)

Do you understand what we represent? We are the promise that the Rosy Cross made to the Immutable. We are the Star Seeds that guarantee the perennial existence of the universe; we are the hand of God that shapes creation. We are the Torch that Christ must bring to the Father to feed the Primordial Fire and reanimate the forces of Life, which, without our contribution, would slowly but surely go out. We hold the key to the universe and must secure its Eternity.
— May 28, 1994, a note found on one of the Solar Temple's computers (Mayer, page 2)

The Critical Events

On October 4, 1994, 53 Solar Temple members were found dead in Switzerland and Canada. They perished in a transcontinental group suicide and homicide to accomplish the lofty purposes alluded to in the above quotation. In December of 1995, 16 more "true solar believers" died in a similar apocalyptic ritual. Five more were added in Quebec in March 1997 (four days before the Heaven's Gate suicides in California). A total of 71 Order of the Solar Temple members have died so far.

The Pipers: Intertwined Psychobiographies

Joseph DiMambro was born in France on August 19, 1924. At age 16 he apprenticed as a watchmaker and jeweler. Even as a teenager he was fascinated with medieval esotericism. In January 1956, he joined the Ancient and Mystical Order Rosae Crusis (AMORC), to which he would belong for twelve years. In the 1960s, DiMambro established relationships with future Order of the Solar Temple figures, such as Jacques Breyer, the leader of "Templar resurgence" in France in 1952. He had some legal problems (swindling and writing bad checks) in the early 1970s. (All Malignant Pied Pipers show various degrees of psychopathy and sociopathy.)

In 1973, however, DiMambro became president of the Center for the Preparation of the New Age, a yoga and mediation school in Geneva, Switzerland. His Golden Way Foundation, financed by a few wealthy members, became the front for the esoteric rites and community of DiMambro's Order of the Solar Temple (or *Ordre du Temple Solaire),* loosely descended from the Rosacrucian Society and the Knights Templars from the days of the Crusades.

As a boy and young man, DiMambro hated his father, who was strict and cruel. DiMambro deeply resented that he had been born in a poor, humble family of Italian immigrant workers. He had higher ambitions and could not forgive his father for having passively accepted the family's poverty.

DiMambro loved his mother deeply and she possibly was the only person he loved in his entire life. She apparently pinned her hopes of overcoming poverty on her son Joseph. In this respect, DiMambro is similar to Jim Jones, whose mother's love was laden with vicariously injected dreams of glory. DiMambro's early narcissistic wounds stemmed from the deep hatred and disappointment he felt toward his father. He also despised and feared the family's poverty. Like Asahara, he departed home for bigger and harsher disappointments.

In 1982, DiMambro, then 58, met Luc Jouret, a 35-year-old homeopathic physician from Belgium. Jouret and his wife were accepted into the Golden Way and took the oath of the Knights of the Rosy Cross on May 30, 1982. DiMambro was aware that he was not a gifted speaker and confided in some members that Jouret had eloquence and charisma; in addition, his professional standing as a physician would give him credibility. Jouret, DiMambro felt, was the type of personality who could be taken seriously as the Order of the Solar Temple expanded.

So DiMambro, the mystical father, would be discreetly backstage, while Jouret, the skilled speaker, was in the limelight of the group's teachings. Jouret was indeed a gifted speaker and drew hundreds of seekers to his charismatic lectures that touted the group's seminars and programs with elaborate and secret initiations, such as the "International Order of Chivalry, Solar Tradition." Jouret was able to strike a reverberating chord with the "seekers" of the time.

In 1984 the Golden Way Foundation paid for the DiMambros to move to Canada, first to Toronto and then to French-speaking Quebec, establishing a second focus of activity for the group. The executive council of the foundation explained the move: "The reason for this decision is simple. North America has become the source of most of the new impulses which determine the way life evolves on this planet. It is therefore fitting that the modern Knight Templar of the old continent should play his part in the Age of Aquarius by adding his inspiration to that which his counterparts in the New World will bring to the planet." (Mayer)

Jouret ran into trouble with Canadian police, however, when he encouraged members of the Solar Temple to buy guns illegally. Part of Jouret's message concerned an ecological apocalypse; he often lamented that Earth was dying of the cancer of pollution, which mankind had created and perpetuated. Members, he said, would need guns to survive and protect themselves as the world's environment deteriorated. The parallels here with Koresh's gun fetish and apocalyptic Biblical prophesies are striking. Jouret's escalating rhetoric of apocalypse and survivalism repelled some members of the cult, who pulled out, but many others remained.

It is no surprise that money and fears of poverty would play a significant role in DiMambro's dark epiphany experiences in his middle age. At the height of its growth in 1989, the Order of the Solar Temple had 442 members. Some large donors to the order hoped to see "life-centers" with survivalists' ideals springing up in Canada and Switzerland. But in the early 1990s, large donors began distancing themselves from the Solar Temple after it was discovered that DiMambro and other leaders were diverting funds to their own travel and living expenses.

DiMambro had pretended since the 1970s that he represented the Mother Lodge and received his orders in a mystical way from masters in Zurich. (Mayer, page 8.) Around 1990, DiMambro's son Elie, then 25 years old, began to doubt the existence of the masters of Zurich and discovered his father's charades and fakery. Young Elie DiMambro, playing a Hamlet-like role in the drama of the Order of the Solar Temple, began to speak openly about how his father faked the illusionary spiritual presences and apparitions appearing in ceremonies of the Order and in their sanctuaries. (Mayer, page 9.)

So DiMambro's mystical illusionary romance with success and money began crumbling from the damage of the psychological dagger wielded by his son (I see here a parallel with Jim Jones' son, John Stoen, as a source and focus of conflict) and the scrutiny of the police. DiMambro's behavior grew progressively more authoritarian, just as his own father's behavior had been. DiMambro began expecting unconditional obedience from members while exploiting rivalries among his underlings.

Just as DiMambro's father had been strict and authoritarian, so was Jouret's. This stemmed from a secular ideal of promoting education, but not from the family's religion. Jouret's father donated many hours to helping children who had troubles with their studies. Yet Jouret's father became physically and psychologically abusive at home while pushing these ideals and promoting "discipline" in his students.

Just as DiMambro had a deep love and affection for his mother, so did Jouret (Mayer, 2001, personal communication). It appears that just as Jouret found an idealized and supportive father in DiMambro,

DiMambro found a loyal and admiring good son in Jouret to compensate for his traitor son, Elie.

Some Malignant Pied Pipers merge with their mother's dreams of glory, such as Jim Jones with his mother, who dreamed that her son would become a famous missionary like Albert Schweitzer. Other Malignant Pied Pipers such as Koresh and Manson sought unconscious revenge toward their mothers who neglected or abandoned them. They accomplished this revenge by gaining power and control over their female followers. The usual family ties and relationships are inevitably perverted, distorted, and used by the destructive cult leaders for their own narcissistic needs in the pursuit of fantasized emotional restitution, revenge, and glory.

In the case of the Solar Temple, there is no evidence that DiMambro or Jouret was sexually exploitive, yet each of them put their fantasies of glory, money, and arsenals of weapons foremost in their lives, exploiting their families, the tenets of the environmental movement, and their followers to feed their delusions.

Dark Epiphanies
Elie DiMambro's criticisms and attempts to demystify his father, and the scrutiny of the Canadian police (including wiretaps) around the Quebec group's arsenal of guns were dark epiphanies that helped propel the cult toward their apocalypse. Rumors and dissent also swept the Swiss cult locations.

Apocalyptic Scenario
The Order of the Temple Solar group death scenario was very similar in its ultimate experience to Applewhite's Heaven's Gate, Manson's Helter Skelter, Jones's People's Temple, and Koresh's Waco.

Jouret and DiMambro drew on the mysticism surrounding the Knights Templar and its mystical brotherhood, which dates back to the 14th century. Just as J.Z. Knight (see Chapter One) claimed an ancient power source, Jouret claimed wisdom and strength from Solar Temple traditions. Jouret wove science into his theology of environmental apocalypse, as he convinced others that the Earth was dying from the cancer of human environmental and social pollution.

As external legal and financial pressures built upon DiMambro and Jouret, the evolving, co-authored scenario of apocalypse grew more grandiose. Just like Jonestown, Waco, and Heaven's Gate, the group death scenario at Solar Temple was "the transit" or "exit from cancerous earth" to bring the "germ of life" to another planet. In the time before the mass suicide, DiMambro grew more paranoid and Jouret grew more depressed.

In December 1993 and January 1994, these leaders began to receive revelations that the planet Jupiter would be the group's new home, just as Jim Jones had ranted in the jungle on special socialist heavens he was preparing, and as Applewhite's group would eventually follow the Hale-Bopp comet.

On the surface, the Order followers showed the usual seeker and in-betweener pattern of passive vulnerability. Solar Temple followers were also heavy on idealism. These searchers were looking for true values that would improve their lives, such as sincerity, humility, wisdom, brotherly love and a genuine direction. Many of the Order of the Solar Temple members were successful, productive, and intelligent people. At the same time, they were also bored with the materialistic and fast-paced shallow world they found themselves in. And for a reason beyond our own guesses, they were drawn to a role that saw them become passive participants in the disillusioned search for new, mysterious, and powerful domains beyond this world.

Diagnosis

I do not have sufficient density of data to apply the DSM-IV criteria and reach a definitive decision about these Malignant Pied Pipers, but they each fit at least 4 of 9 Narcissistic Personality Disorder criteria (grandiosity, preoccupation with fantasy, exploitive, lacks empathy) and probably more.

Even though I cannot document all of the full DSM-IV criteria, their destructive, exploitive cult enterprises qualify them as Malignant Pied Pipers.

In Chapter Six we examine those MPP siren songs and the ears that long for their self-soothing tune.

CHAPTER SIX

THE SIREN SONG OF DESTRUCTIVE CULTS: RECOGNIZING THE MUSIC OF A MALIGNANT PIED PIPER

In my early years of cult study, I assumed that a person lured into a cult must have severe personality weaknesses, problems, or mental illness. I found that this assumption was inaccurate. As we have seen from the biographical accounts of Malignant Pied Pipers, cult followers come from the full spectrum of humanity — young to old, poor to rich, educated to illiterate, conservative to liberal, religious to uncommitted. Anyone can be vulnerable to cult recruitment in certain life circumstances.

If we think of common human needs as a pyramid, the base of the pyramid is built up from the essentials — oxygen, water, food; then clothing, shelter, and protection; and so on in a gradual ascent through community and culture. The fundamental human need to affiliate with small and/or large groups is near the top, just below the domain of spirituality. Spiritual needs are experienced (or denied) individually, and are intensely private and personal. Yet they are also learned, mediated, amplified, and rewarded within a community. All human beings have deep and normal needs to find spiritual meaning in their lives and to affiliate with a group and a community as part of their quest. These aspirations have both rational and irrational elements. (Abraham Maslow, *Motivation and Personality.*)

As we encourage our young people to be spiritually connected with other people, we must remember that there are risks. A wise and mature nurse at our local hospital made the following comment when we were discussing this book: "Dr. Olsson, we raise our kids to be kind, curious, and open to the world and the diversity of peoples' beliefs. The paradox is that this can leave them a little too naïve and trusting, and therefore, vulnerable to clever predators — your MPPs."

Freud observed that affiliation with a group gives power and protection to an individual. Any small or large group forms collective goals, core values, rules, and norms of behavior. Even as the individual is nurtured and supported by the group, he or she often subordinates or compromises individuality in deference to the identity of the group.

Groups require leaders for their formation, administration, and day-to-day operation. Natural leaders generally possess charisma and charm in some degree. Members of the group, in return for investing their own individual power and authority in the charismatic leader, vicariously participate in the leader's power and authority. This idea is critical to understanding the lure of the Malignant Pied Piper. Cult members are not just passive victims of a cult leader's charisma. The relationship involves a powerful co-dependency that resembles a dysfunctional marriage.

Leader-follower relationships in destructive cults are the epitome of codependency, dysfunctionality, and abuse. Membership in a destructive cult is devastating to the individual in terms of his or her creativity, intellectual maturation, and individuation. Destructive and exploitive cult leaders victimize their followers because of their own narcissistic personality problems.

Experts estimate that there are approximately 2,700 cults in America, and they involve as many as three million young people between 18 and 25. Many of these cult groups are absurd or odd, but benign. I estimate that about 100 cults have the potential for a malignant shift that could lead to another Jonestown or Waco. Sometimes it takes decades for the serious suicidal behaviors of a Jim Jones or Marshall Applewhite to gain momentum.

Destructive cult leaders are not benign religious idealists who love people and want to help them. They are spiritual parasites and predators with major personality problems. They need followers to co-author their grandiose dreams and schemes.

All cults offer easy, glib answers to the meaning of life, instant comradeship, and emotional indoctrination called "love-bombing" (Langone, pages 98-99). Exploitive cults offer these things and more. They espouse doctrines and habits that are based on the leader's whims or idiosyncratic revelations, both extensions of his or her personal psychological problems. These doctrines seek to supplant, devalue, distort, or attack traditional religious, family, or community values.

Destructive cult leaders employ, by intuition or calculation, powerful mind-control and recruitment techniques. The cults spend a great deal of time and effort recruiting new members and gaining control over their financial, social, familial, and sexual lives — for their "own good." The exploitive leaders actively seek to alienate cult members from ideas and people in the outside world, including family and community. The cult supplants the family.

An already troubled personality may be particularly vulnerable to cult indoctrination, but any normal person may also be susceptible at certain times. As Dr. Margaret Singer has shown, people are more vulnerable to exploitation by a destructive cult group during periods of change that she calls "in-between times" — in between relationships, divorce and remarriage, high school and college, jobs, and so on. During these times, people often place their trust in the leadership of a spiritual or religious group. I want you to meet a couple I worked with who extended their trust to an exploitive cult leader. It occurred at a time of transition and vulnerability in their lives.

An Important Story of Tom and Tanya, Everyperson
Tom graduated in the top ten percent of his class at a small, prestigious liberal arts college. He was six foot two, powerfully built, with rugged good looks, a charming smile, and a keen sense of humor. He was elected senior class president. He easily got into

Stanford Law School and fell in love with a classmate whose beauty matched her brilliant mind.

Tom and Tanya married after their first year of law school. They felt they could really talk and listen to each other in a way they had not seen their own parents do. One Friday night after supper, Tanya brought up a difficult issue.

"Tom, do you really like law?" asked Tanya.

"We both have a B average," replied Tom from behind the evening paper.

Tanya's voice darkened. "Tom, put down the paper, I am trying to discuss our lives and where our real commitment is. The law and lawyers are boring and myopic. Lawyers aren't taught to care about people — just about power and control."

Tom's stomach tightened as he said, "Love, we can make good money in just a couple of years, and have babies like we have talked about. We could stop accepting 'love money' from our parents and really be out on our own. Besides, with both of our fathers' medical problems … "

Tanya went on. "Tom, remember when we first met, how we would talk about setting up a law center for poor people? We talked about defending abused children and changing rape laws to keep the victim from being attacked in court."

Tom said, "Love, we need to establish ourselves financially first. We have to do 'nuts and bolts' law first, then pursue those ideals." He expected anger or sadness in Tanya's response, but instead she grew detached and misty-eyed and moved beyond what he had just said.

"Tom, have you noticed that I've been missing class on Fridays?"

"Yeah," Tom replied. "I thought you were jogging or visiting with Val and her kids."

"Tom, I've been attending a Bible group called The Class. Reverend Acton conducts The Class like no other class any of us 'freshmen' have ever encountered."

Tom erupted. "Freshmen! Tanya, we're sophomores!"

Tanya smiled for the first time in the conversation. "Tom, you don't understand. Reverend Acton calls us newcomers 'freshmen.'

Will you come to The Class with me next Friday? The people there are so friendly and caring."

"What does he talk about? What is expected?" Tom asked.

Tanya said, "Honey, I'd like you to hear for yourself. We can talk afterward — the 'Upperclassmen' often take us out for coffee and doughnuts after The Class."

As Friday approached, Tom grew vaguely uneasy, almost in direct proportion to Tanya's enthusiasm for The Class. Tom noticed Tanya was reading handouts from The Class rather than studying her law books. She refused to let Tom see the handouts and told him to be patient until Friday.

On Thursday, Tom skipped school for the first time since starting his study of law. He tried to call his father in Chicago but couldn't reach him. He decided to drive to Golden Gate Park and take a long walk. Tom felt mild guilt over missing class, but soon relaxed in the cool mist.

Suddenly a siren blared in his ears and an ambulance flew past on its way to the hospital. Tom got a brief but vivid glimpse of the patient inside. The man was balding, ashen-faced, and his features were twisted with pain and terror. Tom thought instantly of his dad, although he knew his father was in Chicago.

Then Tom began to think about his own death. Once in college he had read a book by Margaret Craven called *I Heard the Owl Call My Name,* about an aging priest who died while serving a remote Indian village. Tom's professor gave him an A on his review and remarked that he liked Tom's term for death — "the great pregnant silence beyond all words and worries." Tom remembered a statement in another book he read in the course — "It is not Death that men fear the most, it is death without some sense of personal significance," from Ernst Becker's *The Denial of Death.*

Tom thought about the way in which death and the meaning of life were entwined in a tangled skein. Maybe Tanya was right. They needed deeper meaning in their lives beyond the endless details of legal texts and the visions of future salaries dancing in their heads.

The Class met promptly at 9:30 a.m. and Tom and Tanya were there. No one introduced the Rev. Acton – he suddenly appeared at the podium and began moving toward the class. He had jet-black hair, thick eyebrows, and an almost copper-colored skin. His stare was penetrating, yet gentle and seductive. He had a lean but muscular build, with catlike movements. He wore lightly tinted glasses, white pants, and a dark blue shirt that shimmered.

Rev. Acton said, "We have two new potential freshmen here this morning – who will introduce them?"

A male voice behind Tom said, "Reverend Acton, Sharon is right in front of you. I will let Tanya introduce her husband Tom."

Rev. Acton spoke directly to Sharon. "Did you know that your name stirs allusions in the scriptures to roses and other gentle flowers?"

Sharon's face showed a mixture of surprise and veiled pleasure.

Tom thought, What a phony charmer.

Before Tanya could speak, Rev. Acton whirled to face Tom, saying, "You are Thomas — do you know about Doubting Thomas? He was probably a lawyer like you are studying to become. Doubting Tom insisted on seeing and feeling the resurrected Christ's scars before he would believe in Him."

Tom was sweating profusely and tongue-tied. He turned to Tanya, but she was smiling at Rev. Acton, transfixed by his charisma.

Rev. Acton moved on quickly, saying, "I respect your profession and your confident skepticism, Tom." He gave Tom a brief hug and started the morning's lesson.

Tom felt blown away. He gradually identified feelings of repulsion, but also intense pleasure at being hugged. His mother or father had never hugged him that way. He felt bombarded with warmth and a strange, defiant delight.

Rev. Acton talked, his voice rising in a steady crescendo, about "the joy of devoting oneself to others in the community" and "the peace attained when one's vanity is overcome by helping others." Tom didn't hear much detail because he was focused on the warm spots on his shoulder, arms, and chest where Rev. Acton had touched him. He felt he had failed in this, his first day in court. Yet he felt strangely loved by this charismatic, self-appointed man of God and Goodness.

Rev. Acton's parting words haunted Tom.

"Tom, don't be afraid of your goodness."

Five Years Later

Tom worked several evenings in The Class's soup kitchen and clothing center. Tanya taught reading to inner city kids at the parish school across from Rev. Acton's office. Rev. Acton had raised money to pay for Tom and Tanya's California Bar review course, and he put on a Class ice-cream social in their honor when they passed the bar exam. He was raising funds now for a legal clinic for indigent clients. Tom and Tanya found "exhausting joy," as Rev. Acton called it, in their 12-hour days working for the Class.

One evening the soup kitchen closed early. That was unusual, but Tom didn't mind – he was tired and had not seen much of Tanya for a week. She had seemed distant lately, and he looked forward to talking with her. He realized he had forgotten his key and went to the back door of their small duplex, hoping to see Tanya in the kitchen. Suddenly he heard moans from inside. Tom glanced into the bedroom window through a slit under the shade. Rev. Acton was above Tanya, thrusting in passion, grunting, "Oh Tanya, this child will lead some fine Classes some day!"

Tom's head swam and he fell hard against the house, hitting his head. He awoke with Acton holding him and Tanya pressing a cold, wet towel against his forehead.

Tom said, "I can't see. I'm blind! My head is killing me."

Rev. Acton calmly said, "We'll take him to the hospital in my car."

In the emergency room, a physician checked Tom over but was puzzled about his blindness. There was only a slight abrasion on Tom's head and no skull fracture. A neurologist found no sign of neurological disorder. The consulting psychiatrist, Dr. Brown, confirmed a diagnosis of Psychogenic or Hysterical Blindness and transferred Tom to the psychiatric unit.

In total darkness, Tom felt terrified and alone. It was hard to tell the blackness of sleep from the blackness of his blindness. During restless sleep, he dreamed about looking at family photo albums and pictures from his wedding to Tanya, but in the dream Dr. Acton hid the albums and smiled in a strange, sly way.

Tom awoke to Dr. Brown's soothing voice. "Tom, let your muscles relax and keep your eyes closed. Now, as I count to ten, you will gradually get sleepier and sleepier, but also you will be relaxed and clear-minded, like in a pleasant but vivid dream. You are perfectly safe. I will be with you the whole time. 1-2-3-4-5-6-7-8-9-10. Tom, we are outside your house last night. Are you with me?"

Tom heard himself say in a clear voice, "Yes, Doctor Brown, but I forgot my key. We can go around to the back. Tanya's probably in the kitchen."

"Okay, Tom, I'm with you."

Suddenly Tom asked, "What's that moaning sound?"

Dr. Brown replied firmly, "What do you *see*, Tom?"

"My God, it's him! Them! Stop!!" Tom's whimpers turned to long, gulping sobs for several minutes.

Then Dr. Brown's gentle, firm voice said, "Tom, I'm going to count backward from ten, and when you return, you will see and remember all that happened last evening. You will be able to see clearly, and we will talk about it. Ten –9-8-7-6-5-4-3-2-1 – Tom, I'm here."

Tom exclaimed, "I can see! But I wish I didn't."

Tom cried and talked with Brown for a long time. He saw Dr. Brown three times a week for many months. Tanya attended one of their sessions. She freely admitted her "spiritual and sexual relationship" with Rev. Acton, explaining it as a special spiritual domain of opportunity and no threat to their marriage. She said that Rev. Acton had chosen several members of The Class to commune with sexually. Rev. Acton preached that in The Class, a new and transcendent loving consciousness brought them all beyond the confines of "the prison of traditional marriage."

Tom erupted. "Tanya, what about our wedding vows?"

Tanya rolled her eyes and stared with a vacant half-smile as she said, "Rev. Acton says, and I agree, that many of us in The Class have grown beyond our 'family values of origin.'"

Dr. Brown quietly asked, "Tanya, had you been talking to Tom about these 'value shifts'"?

Tanya replied, "I tried and so did Rev. Acton."

Tom shouted, "Damn Acton to hell, that slimy psychopathic hypocrite!"

Tanya responded resolutely, "Tom! Take that back — now!"

Tom said, "Hell no, Tanya. Can't you see Acton for the manipulator he is?"

Tanya said with cool detachment, "Tom, grow up. Rev. Acton is the father of my child — or should I say, The Class's blessing child."

Dr. Brown breathed deeply. Tom grew pale and speechless. Tanya walked out.

The divorce was painful for Tom and took many months. The Class tried to get alimony and child support from Tom, but when he demanded genetic testing and threatened a paternity suit, Tanya quickly agreed to a no-fault divorce.

The Class is no innocent alternative religious group. It is a destructive and exploitive cult. Could the cult end in apocalypse? Only time, the dynamics of Rev. Acton's character disorder, and the Class members' response will tell.

The Four Core Cult Indoctrination Techniques

In addition to the "love bombing" described earlier, exploitive cults use four basic processes, described by Dr. Margaret Singer, to indoctrinate, control, and retain members.

1. Resocialization. The new recruit is deluged with a whole new social network. Contact with the member's family, church, and community of origin is discouraged as the social network and duties of the cult become primary. Television and print media are withheld, tightly censored, or reinterpreted by the leader.

2. Reparenting. The cult leader and his or her key lieutenants are foisted on the recruit as authoritative parental figures. The leader often insists on being called "father" or "dad." Biological parents are designated as "merely flesh relationships" or "suppressives." Mail and phone calls from family and friends are discouraged or actually prevented.

3. Behavior Reconstruction. The cult has strict guidelines for sexual, social, vocational, financial, marital, and daily behavior. The cult leadership controls pairings, reconstitutes families, and dictates sexual practices.

4. Self-reconstruction. The recruit is treated like a new person whose role and position in the cult community is assigned and dictated by the leader, putting the new member in a position of powerlessness and passivity. The recruit may even be assigned a new name.

Typical Tactics of Destructive Cults

In order to break the will of new recruits and mold them to the cult, leaders use effective mind-control tactics.

* Housing, meals, and even bathroom breaks are controlled by the leader during lengthy cult lectures.

* Rigid, repetitive group confessionals, filled with over-simplified rules and clichés, work on the minds of members in a context of sleep deprivation and rhythmic music and chanting.

* Cult recruiters learn to use their eyes aggressively, staring intently two inches behind a recruit's head or forehead while smiling steadily and repeating cult dogma hypnotically.

The Profound Effects of Cult "Mind Games" on the Recruit
Unlike life experiences such as religious education or other maturational passages, the cult mind-control techniques and tactics described above cause profound mental changes in previously intelligent and healthy people. Recruits experience reduced cognitive capacities and mental flexibility. The cult methods cause blunted, constricted emotions that lead eventually towards a regressive, naïve, and childlike dependency. Cult victims often experience psychopathological symptoms such as hallucinations, illusions, and severe psychological detachment or disassociation. Cult victims frequently lose weight and affect a falsely bright expression that masks their depression and disassociation. These changes in mental status, often shocking to victims' family and friends, can persist for nine to 18 months after leaving a cult, even in a formerly strong and resourceful person.

In therapy, a recovering cult member will often idealize and depend upon his or her psychotherapist, but a good therapist will help the patient to regain his or her confidence and self-reliance. In contrast, the exploitive cult leader tries to use and exploit a cult member's idealization and dependency in perpetuity.

The Story of Cindy: Cult Seduction Strikes a Normal American Family
Cindy was 19 and a sophomore in college when her parents sought a consultation with me. Cindy had been quite homesick during her freshman year, but now she was making a few friends, her grades were good, and she had a part-time job at the student union. She had decided to stay at school during her spring break to finish a term paper and save the airfare home.

Her parents grew mildly concerned when they heard nothing from Cindy during her break. Finally they reached a roommate who said she was worried about Cindy. She explained that Cindy had joined a group called The Family and was going to nightly lectures and weekend group retreats. The roommate said, "Cindy has lost weight and acts real spacey sometimes. I hear that The Family is pretty weird. They say that associating with your own family and their church is 'like eating your own vomit.' Weird."

Cindy refused to talk to her parents on the phone, telling her roommate to relay the message that she was independent now and was very happy with her new projects. Her parents grew more concerned when they learned that Cindy had quit her job and moved into a "free" dorm owned by The Family. Her grades dropped to Cs and Ds. Her new dorm counselor rarely let her speak to her parents. In a brief conversation, they learned that she was working at The Family's recycling center and was gone every weekend. Her parents thought Cindy sounded tired, detached, aloof, and both vaguely hostile and inappropriately cheery. Then they got a terse letter from Cindy telling them she was staying at school for the summer.

I advised Cindy's father to visit her immediately. He was alarmed to see that Cindy had lost a lot of weight — about 20 pounds — and had a peculiar detached but smiling expression. She stared at her dad's forehead and talked about "Universal Christian love sources," and proudly reported being selected to cook for and babysit for The Family's leader, Brother Smith. Cindy's father was not permitted to visit with her alone. Counselors were always with them, as well as a boyfriend that Brother Smith had especially selected for Cindy. The cult members frequently chanted, sang "Family songs," and danced or rocked rhythmically back and forth.

When Cindy's father called the cult office, he learned they had no board of directors, no bishop, no records, and only a "spiritual leadership." He was told his daughter had freely joined The Family and as an outsider he should have no concerns. He talked to Cindy's former roommate, who told him that Cindy and others went to group "confessions" where they admitted to "sins of the flesh," danced, sang, and participated in group trances until all hours. Cindy and her boyfriend slept on mats at the cult office so they could do cult work for Brother Smith at any time.

When her dad confronted Cindy, she spurned him, saying, "We are only flesh relations now — not family any more." Her boyfriend called him a "suppressive" and asked him to leave. Cindy refused to go home with her father and a young cult attorney advised her father not to interfere with her civil rights.

Cindy's father returned home a shaken man. We had an emergency family meeting at which Cindy's older brother came up with a plan. Cindy's mother called her and lied. She told Cindy that her beloved grandmother was very ill and wanted to see her. At first Cindy refused, but then changed her mind when she was told that her grandmother was near death.

When Cindy's flight arrived, a large group of her high school and church friends and her family were there to greet her. Her cult boyfriend and a "counselor" who had accompanied Cindy were told they were not welcome. Cindy cried and her boyfriend protested, but he could not prevail and retreated to a telephone to call the cult.

Cindy was naturally angry and confused when she learned that we had tricked her into coming home. At my recommendation, at least three friends or family members stayed with her at all times for the first three months. We had two family therapy sessions each weekend for two months, and I saw Cindy twice weekly for individual therapy.

With my support, the family was able to listen in great detail to the astounding story of Cindy's indoctrination into The Family. They were careful not to attack the cult that she had loved and idealized. At her individual sessions she alternated between rage at me and awe of me. I helped her see this transference reaction toward me as a reflection of her true feelings toward Brother Smith. Soon she was able to tell me how much I resembled him. Brother Smith had a red beard like mine and he was gentle, persuasive, and charismatic, with an amazing and detailed knowledge of the Bible. I included Cindy's clergyman in several sessions and he was helpful in the process of her therapy because he could calmly confront Smith's distortions of Christian theology.

Cindy required antidepressant medication for the first nine months of therapy. Two years later, after twice-weekly therapy and monthly family sessions, Cindy had been able to grieve her loss of the gratifying balm of the cult. She still would occasionally "trance out" and hear the cult hymns, mantras, and feel the rhythmic dances. On one occasion toward the end of the first year, Cindy's cult boyfriend and Brother Smith called when she was home alone. She nearly agreed to return to them. Using guilt and shame tactics, Brother Smith told her that she had psychologically traumatized the children in her class at The Family's school by leaving so abruptly. Fortunately, Cindy was able to stall long enough to tell her family and me about the cult pressure. With our support, she was able to resist.

Saving Cindy

The cult used "love bombing" to recruit Cindy and shaming to try to retain her. Fortunately for Cindy, she had the help of a truly loving, tenacious, and assertive family. She also had psychotherapy that was informed about destructive cults and their tactics.

Cindy initially experienced The Family as her own new and special affiliation that helped her move away from her dependency on Mom and Dad. She was a classic example of Dr. Margaret Singer's "in-betweeners" who are vulnerable to cult recruitment as they search for affiliation with a group that seems to have a good cause and ideals.

Cindy related that when the cult recruiters first approached her she had just broken up with a boyfriend. She was far from home and felt lonely between semesters. She had just finished work at the student union and was standing alone, looking at the bulletin board, when several cult members approached her and pointed out the notice for their "Philosophy discussion group" that evening. They all expressed an active, effusive interest in her and aggressively urged her to attend their meeting.

Cindy described this first encounter: "They smiled brightly and stared into or behind my eyes — like into the middle of my forehead. One guy had a guitar and sang a special 'sweet song for Cindy.' At the time I didn't realize these 'music men' meant 'trouble in River City.'"

She told me that once she began attending meetings of The Family regularly, there was a subtle but progressive shift in how the cult approached her. She was first encouraged, then expected, to "use my creativity" on a recruitment team to get new members. She was instructed to find others to join "our special, loving Family."

Prevention and Treatment
The best prevention against becoming a cult victim lies in the hard psychological work required in building good communication in your marriage and family. Be aware of "in between" times for yourself and your family members. The most difficult times for families are divorces or family moves. Pay the piper of quality time spent listening to each other.

Work at building close friendships and being a good friend. Family friends are particularly valuable for an "in-betweener." As a young person is separating from the normal dependence on Mom and Dad, they often will be more open to hearing Uncle Bill's ideas about decision-making, while ignoring or tuning out their parents. If a friend's son or daughter is home for the summer or between semesters or between relationships, take them out to lunch and listen. Spend time, listen, and tune in.

Invest time and money in church and community activities and projects. If you attend church or synagogue, take an active part in its activities. We all hunger for healthy affiliations to help provide meaning in our lives. No person is a "Robinson Crusoe" without a person "Friday" or a group with which to share ideals and mutual support.

Beware of and actively challenge any person who is rapidly engaging and takes an extraordinary, aggressive, or unusual spiritual interest in you. Especially beware if they seek extensive information about you and share little about themselves or their motives.

Treatment of a person who is ensconced in a cult is notoriously difficult. Deprogrammers provide a treatment that can be worse than the cult disease.

Here are a few techniques that can be effective if a friend or member of your family has been drawn into a cult:

*Work tenaciously to keep the continuity of the relationship.

*Keep sending letters, news clippings from home papers, and pictures of home and family. Encourage friends and relatives of the cult victim to do the same.

*Consult with trusted friends and lawyers, clergy, and therapists who have knowledge and experience about cults.

*Personally attend cult speeches and sermons. Take an interest in the cult's ideas and ask rigorous but respectful questions.

* Actively inquire about literature and policies of the cult. Ask about their financial records and policies.

When a family member leaves a cult:

*Give them active love and support.

*Get individual and family therapy.

*Make sure the therapist is well-informed about the post-traumatic effects of cult practices and mind-altering techniques.

Knowledge and awareness about Malignant Pied Pipers can save your family's soul.

The Ultimate: Terror Cults and Mind-Control
The scope of recruitment in terror cults is on an exponentially grander scale than the recruitment of members to cults like those started by Jim Jones or David Koresh, but they all are built on the same interweaving of theology and the "just cause" by the self-

appointed messiah. Radical Islamic terrorist groups like bin Laden's Al Qaeda use and twist Muslim teachings to their own ends in recruiting members in Arab and Muslim countries and training them in *Madrassahs* (religious schools) or terrorist camps.

Recruits to terrorist cults come to see death by martyrdom as a desirable end that brings respect to the martyr's family. Idealistic and naïve children are recruited and trained in these camps using many of the same core techniques as other destructive cults.

In her profoundly important book *Terror in the Name of God: Why Religious Militants Kill* (Harper Collins, 2003), author Jessica Stern describes in detail *Madrassahs* in Pakistan. She offers a poignant interview with Syed Qurban Hussain, a traditional doctor of herbal medicine and father of seven sons, all of whom were trained as *mujadhadeen* fighters in Afganistan. One son died a martyr in the *jihad*. He had been educated through the eighth grade, learning the Koran by heart in a *Madrassah*.

Stern writes, "Hussain tells me he is happy to have donated a son to the cause of *jihad*. Whoever gives his life in the way of Allah lives forever and earns a place in heaven for seventy members of his family, to be selected by the martyr. Everyone treats me with more respect now that I have a martyred son. And when there is a martyr in the village, it encourages more children to join the *jihad*. It raises the spirit of the entire village."

The Al Qaeda genies are already out of their bottles all over the world and have a virtually unlimited supply of recruits in the *Madrassahs*. The only solution I can see to this Ultimate Malignant Pied Piper operation is a revamping of our U.S. foreign aid policies so that brutal dictators in Arab and Muslim countries don't get our money — but the people who really need economic help and broader educational options than the *Madrassah* do get the help. We also need to expand the Peace Corps and spend less money on "smart" bombs, missiles, and bullets. Muslims need to see Americans smiling, teaching, helping, and listening, and not looking through the sights of rifles.

The ultimate Malignant Pied Piper awaits our scrutiny in Chapter Seven. Hold on to your headscarves!

CHAPTER SEVEN

THE MIND OF OSAMA BIN LADEN:
THE ULTIMATE MALIGNANT PIED PIPER?

On September 11, 2001, my first patient of the day mentioned in passing that a plane had crashed into the World Trade Center Tower in New York City. My second patient said that a second jet had crashed into the Twin Towers. Then I knew that it had been no accident.

After a week of going through periods of shock, fear, and rage, I wrote a poem, as I often do as a form of therapy.

Eagle Wounded (9-11-01)
Bizarre flying Trojan Horses. Souls engulfed by fire.
"Pilots, give them what they want and we are safe."
My God! throats of fragile freedom cut, painfully.
Horror decades envisioned for angry revenge seeking.
Listening now paralyzed by greedy gulps of hatred.
Forgiveness blurred in scared scenes scarred by loss.

Eagle Healing (9-19-01)
Searching tirelessly in the rubble, with love-tears.
Transforming pain and rage, with healing prayer.
Light of ideals, smashing Hatred's endless night.

128

Destruction's curse harnessed, by long-listening.
The mourning, then a new world morning song.
Healing Eagle reaching out, embracing new friends.

In the midst of my anger at Osama bin Laden, I asked myself: Who is this man bin Laden? Why did he want to kill Americans? Why did some Muslims and Arabs cheer and rejoice when they saw the World Trade Center and Pentagon's flames of destruction?

I began to apply some of my 20-plus years of research into destructive and apocalyptic cults and their leaders (Malignant Pied Pipers) to bin Laden and his Al Qaeda network of cult-like terror groups. I considered bin Laden's cyberspace fundraising and recruitment, his charismatic, religion-flavored recruitment techniques, and his disdain for his Saudi homeland community and America as having some similarities to exploitive cults. Osama bin Laden reminded me of destructive cult leaders who harbor childhood resentments about their childhood disappointments and seek to alienate their followers from their families and countries of origin so they can become the great leader-parent figure.

Like other Malignant Pied Pipers, bin Laden's outlook is paranoid and apocalyptic. Paul Berman, writing in *The New York Times,* describes the Al Qaeda terror cult in similar terms: "Al Qaeda upholds a paranoid and apocalyptic worldview, according to which 'crusaders and Zionists' have been conspiring for centuries to destroy Islam. And this worldview turns out to be widely accepted in many places ... a worldview that allowed many millions of people to regard the 9/11 attacks as an Israeli conspiracy, or perhaps a C.I.A. conspiracy, to undo Islam. Bin Laden's soulful, bearded face peers out from T-shirts and posters in a number of countries, quite as if he were the new Che Guevara, the mythic righter of cosmic wrongs." *(The New York Times*, March 23, 2003.)

As I learned more biographical details about Osama bin Laden's childhood and early family history, I noticed some similar patterns to the psychological dynamics and character patterns that I had discovered in destructive cult leaders. The available psycho-

129

biographical information about Osama bin Laden is incomplete and contradictory in many areas, but I think there is enough data to form psychoanalytic hunches and some solid conclusions about bin Laden's personality and his psychological agendas.

It is so important to try to understand the uniquely chilling social-psychological fit between the religiously flavored charismatic leadership of a man like Osama, and the group psychology of communities where he and his colleagues and mentors recruit devoted terrorists. In my opinion, it is a serious mistake to glibly label and dismiss bin Laden as simply a mass murderer, thug, or criminal. In painful truth, he is a Robin Hood-like figure and a spiritual Pied Piper hero for many people in the Muslim world. As ancient martial arts and warrior codes point out: 1. Hasty declaration of war inflates the status and grandiosity of the enemy; 2. Contempt fuels the enemy's outrage; and 3. Incorrect and dangerous assumptions are easily made about an enemy one has only contempt for, and scant knowledge about.

In essence, "If you know the enemy and know yourself, you need not fear the result of a hundred battles. If you know yourself, but not the enemy, for every victory gained you will also suffer a defeat." (Sun Tzu, *The Art of War*, 500 B.C.). This chapter is an effort to know our enemy, Osama bin Laden, from a psychobiographical and social psychological perspective.

I also detected in bin Laden's life trajectory evidence of what I have called Dark Epiphanies in destructive cult leaders. These later life experiences reify and magnify their earlier molding experiences of disappointment, neglect, shame, and humiliation influenced by parents and other childhood role models. In adolescent or young adult life phases, antiheroes are often chosen to counteract disappointment or humiliation/shame experiences with parental figures.

Like other Malignant Pied Pipers, bin Laden's appeal has a unique fit for normal adolescent rebelliousness. Anna Freud said of adolescents, "On the one hand, they throw themselves enthusiastically into the life of the community, and on the other, they have an overpowering longing for solitude. They oscillate between blind submission to some self-chosen leader and defiant rebellion against any and every authority. They are selfish and materially minded and at the same time full of lofty idealism." (Anna Freud, *The Ego and the Mechanisms of Defense*, 1936, pages 137-138.)

What would be normal adolescent rebellion and protest for these young people becomes terrorist actions under Osama bin Laden's tutelage. The Arab world's turmoil creates many young adults who are in the phase of what psychoanalysts call "prolonged adolescence." Siegfried Bernfeld described a specific kind of male adolescent development called "the protracted type." It extends far beyond the usual time frame of adolescent characteristics and is conspicuous by "tendencies toward productivity, whether artistic, literary, or scientific, and by a strong bend toward idealistic aims and spiritual values and activities." (Siegfried Bernfeld (1923), "Uber eine typische Form de männlichen Pubertät," *Image*, IX.)

There are uncanny similarities in "follower psychology" between Al Qaeda recruits and recruits in many destructive cults. The followers are not all poor or uneducated young people. Bin Laden's lieutenants and Al Qaeda leader colleagues are mostly educated and dedicated to his ideal of *jihad*. Radical Islamists like bin Laden recruit by using personal charisma and manipulating the Koran to form attractive mixtures of theology turned into starkly articulated radical political action ideology.

Al Qaeda means "the Base" in Arabic. In addition to enlisting well-educated youth, radical Islamists also recruit poor and less-educated Muslim foot soldiers through religious *Madrassah* schools and young-adult mosque programs and activities. But Osama's *jihad* appeals particularly to disaffected Arab and Muslim youth. These adolescents and young adults have suffered while watching their parents being humiliated, oppressed, and impoverished in countries ruled by wealthy dictators who are perceived as puppets of America. The *Madrassah*-type "schools" offer economic advantages and spiritual inspiration to families and Muslim communities that have few alternatives.

The recruiting techniques of Al Qaeda and its metastatic subsidiaries are clever, creative, and diverse in their applied theology. This element of Al Qaeda recruitment is very different from that of other destructive cults I describe. Exploitive cults in the West seek to alienate young people and older recruits from their families and communities of origin. Al Qaeda, often through the *Madrassahs,* woos local Muslim families and communities for its own ends. Especially in Pakistan, many *Madrassah* students and their families see membership in Al Qaeda and participation in *jihad* as a high calling.

The 9/11 Martyr Pilots: Young Adult "In-Betweeners" in a Prolonged Adolescence

Most of the 9/11 suicide pilots were in their 20s or early 30s. Young adults, either because of prolonged adolescence and/or extreme religious devotion, can be recruited to terrorist groups. In Al Qaeda training camps, they not only teach bomb-making skills, weapons training, and physical fitness, but they also prepare the souls of future terrorists via charismatically presented radical Islamist theology. Radical Islamist clerics preach their distorted version of the Koran in radical *Madrassahs* in Pakistan and some mosques in Germany, Belgium, France, and even Newark, New Jersey. In the long run of this new century, bombastic radical Islamist preaching poses as grave a danger to Western civilization as bullets and bombs. Late adolescents love these exciting, idealistic causes and outlets.

Writing in *The New York Times,* Thomas Friedman described a typical biographical profile for the likes of the 9/11 suicide pilots (Mohammed Atta, 33, and Marwan al-Shehhi, 23): "It's the same story: He grew up in a middle-class family in the Arab world, was educated, went to Europe for more studies, lived on the fringes of a European society (Germany, Belgium, etc.), gravitated to a local prayer group or mosque, became radicalized there by Islamist clerics."

Friedman continues: "Here's the truth: what radicalized the September 11 terrorists was not that they suffered from poverty of food, it was that they suffered from a poverty of dignity. Frustrated by the low standing of Muslim countries in the world, compared with Europe or the United States, and the low standing in which they were personally held where they were living, they were easy pickings for militant preachers who knew how to direct their rage." (Thomas L. Friedman, *The New York Times* Op-Ed page, Jan. 27, 2002.)

Is Osama bin Laden the Ultimate Malignant Pied Piper?
As described above, he is in many respects the ultimate Malignant Pied Piper. How does he differ from the other cult leaders we've described?

Similarities and Important Differences Between Osama's Terrorist Network and Destructive Cults

1. Osama bin Laden does not berate his followers or potential recruits. He does not use "love-bombardment" like Western cult leaders. Rather, he uses a quiet, devout charisma and the hero myths that have built up around his heroism during the defeat of the Soviets in Afghanistan. Here is an example of his prose, as broadcast on Al Jazeera satellite television on October 6, 2002: "I am telling you, and God is my witness, whether America escalates or de-escalates this conflict, we will reply to it in kind, God willing. God is my witness, the youth of Islam are preparing things that will fill your hearts with tears. They will target the key sectors of your economy until you stop your injustice and aggression or until the short-lived of the U.S. die." (Quoted by Rohan Gunaratna in *Inside Al Qaeda: Global Network of Terror,* page xvii.)

2. He does not assert overt control of his members' sexual and financial lives, although his austere camps and strict Muslim codes tend, like those of the Taliban, to be chauvinistic and constrictive. In Afghanistan, according to

Peter Bergen, "The Taliban had imposed their ultra-purist vision of Islam on much of the country." This included banning women from using cosmetics; requiring women to wear the all-enveloping burqa and stay at home unless accompanied by a male relative; forbidding men to shave or trim their beards; and banning smoking. (Peter Bergen, *Holy War, Inc.*, page 158.)

3. Though bin Laden does not attack followers' nuclear families, his harsh attitude toward "infidels" (Muslim, Christian, or Jew) fits the murderous, bloody Wahhabi fundamentalist ethos. For all Muslims, it is "Osama's way or the highway." In *Blind Trust,* Vamik Volkan defines this ethos: "The term Wahhabism comes from the name of Mahammed Abdul Wahhab (1703-1792), who founded an ultra-traditional Islamic cult in Arabia in 1740, near where Riyadh now stands. Wahhabism demands extreme punishments, including execution, for sexual transgressions, drinking, and listening to music." (Vamik Volkan, *Blind Trust,* page 300, note #341.) After 1800, Wahhabis murdered those who opposed them. Ibn Saud later made Wahhabism the official religion of Saudi Arabia. Osama bin Laden is a Wahhabi.

4. Bin Laden's shadowy network does not sharply define "in-group, out-group" boundaries like the Western cults I describe. He seeks to establish a worldwide fascist-Islamist utopia of the Caliphate. I use the term "Islamist" here to indicate the most radical Muslims. "Islamists call the unified pan-Islamic state that rules the entire hub of Islam, and ultimately the entire Muslim world, the Khilafah (Caliphate)." (Bodansky, page 19.) This "state" is arrived at by *jihad,* an armed struggle. I consider this armed struggle to be a world war.

Roham Gunaratna explains Al Qaeda's operational strategy in these terms: "As defined by Osama, Al Qaeda has short-, mid-, and long-term strategies. Before 9/11, its *immediate* goal was the withdrawal of U.S. troops from Saudi Arabia and the creation there of a caliphate. Its *mid-term* strategy was the ouster of the 'apostate rulers' of the Arabian Peninsula and thereafter the Middle East, and the creation of true Islamic states. And the *long-term* strategy was to build a formidable array of Islamic states — including ones with nuclear capability — to wage war on the U.S. and its allies."

Listen to Osama in an August 1996 edict from Afghanistan (*fatwa* means a decree issued by a religious leader): "It should not be hidden from you that the people of Islam have suffered from aggression, iniquity, and injustices imposed on them by the Zionist-Crusader alliance and their collaborators to the extent that the Muslims' blood has become the cheapest in the eyes of the 'world', and their wealth has become as loot, in the hands of their enemies." (Gunaratna, page 119-120.)

Osama then cites as evidence the "massacres" in Palestine, Iraq, Lebanon, Tajikistan, Burma, Kashmir, Assam, Phillipines, Fatani Ogaden, Somalia, Eritrea, Chechnya, and Bosnia Herzegovina. Osama grandiosely insists that all of these troubled locations fit his pronouncements.

Bin Laden continues to extend his grandiose diatribe: "... all of this — and the world watched and heard and not only did they NOT respond to the atrocities, but also, under a clear conspiracy — between the USA and its allies, under the cover of the iniquitous 'United Nations' — the dispossessed people were even prevented from obtaining arms to defend themselves. The people of Islam awakened, and realized that they were the main target for the aggression of the Zionist-Crusader alliance. And all the false claims and propaganda about 'human rights' were hammered down and exposed for what they were, by the massacres that had taken place against Muslims in every part of the world." (Gunaratna, page 120.)

5. Many authorities feared that even after the deaths of Jones, Koresh, Applewhite, Jouret-DiMambro, or the imprisonment of Manson and his Family, their sinister activities would linger on. Most experts agree that even when bin Laden is gone, Al Qaeda and its collegial network of terror groups will certainly live on.

In Search of a Father/Himself: A Psychobiography of Osama bin Laden

Osama bin Laden's father, Mohammad, born in about 1930, immigrated as a laborer from a poor Yemeni family to Saudi Arabia in the late 1950s. His son Osama was born in Saudi Arabia on March 10, 1957, in Riyadh. Osama means "Young Lion" in Arabic. Mohammad bin Laden became a skilled engineer, started a construction company, and gained the enduring respect and favor of both King Saud and his successor, King Faisal. Mohammad was a trusted confidante of King Saud. King Faisal gave Mohammad bin Laden the contract to rebuild the Islamic holy sites at Mecca and Medina in 1973.

Mohammad bin Laden had eleven wives during his lifetime; Osama has more than 50 siblings. Osama's mother, Hamida, is Syrian, and Osama is the only son of her marriage to Mohammed bin Laden. Hamida refused to accept the traditionally passive female role in the marriage. Sharp conflicts over this issue led Mohammed to banish Hamida to another town (Tabuk). Osama, already of low family rank as the 17th son, thus lost his mother, because she lived in a separate household. Osama was raised by his stepmother Al Khalifa, another woman of strong personality for whom he eventually held respect and affection, despite the loss of his real mother on a day-to-day basis. (Robinson, pages 39-40.)

Mohammad spent his time making hundreds of millions of dollars for his Bin Laden Corporation. Some quality time for Osama and his father occurred during an annual weeklong "male-bonding" winter hunting vacation in wild desert regions of the Saudi Kingdom. Osama blossomed during these brief trips with his father and his royal friends. Mohammed was very impressed with Osama's prowess in the desert, and probably recalled his own boyhood adventures in the Yemeni desert. Osama apparently out-shone his less adventurous half-siblings. Osama also excelled in Islamic studies, which drew positive attention and praise from his father when Osama was a boy. Sadly, this "Piper-time" was far too brief and I agree with Robinson's implication that it left a profound father-hunger in Osama's heart. (Robinson, pages 50 and 55.)

Osama bin Laden's Father's Death as Psychological Wound and Narcissistic Injury

Though a multimillionaire, Mohammad bin Laden was poverty stricken when it came to paying the Piper of parenthood. It was impossible for him to spend quality time with each of 50 children, particularly his bright and perceptive son of the "lesser" Syrian wife. He espoused religious piety but worshipped money. Osama was only 10 when his father disappeared from his life. Mohammad bin Laden was traveling home when his helicopter crashed in the desert. Observers reported that 10,000 men gathered for his funeral. Osama seemed deeply affected. Robinson says: " … his grief was deeper than simply the loss of a loved one. Beneath the surface, he had long repressed a deep gouge in his psyche caused by the partial loss of his mother (via divorce) and a relationship with his father shared with so many siblings, a handful of wives, and the pressure of a vast business empire." (Robinson, pages 54-55.)

Robinson continues: "Before his [father's] tragic death, his [Osama's] interest in Islam had drawn him closer to his father. It was a paternal relationship he [Osama] craved, yet Mohammad's sudden death had robbed the youngster of a chance to enjoy anything other than fleeting moments of satisfaction. Family members recall him reeling emotionally." For months, the boy withdrew into himself.

Sigmund Freud described his reaction to his own father's death: "It was, I found, a portion of my own self-analysis, my reaction to my father's death — that is to say, to the most important event, the most poignant loss, of a man's life. Having discovered that this was so, I felt unable to obliterate the traces of the experience." (Sigmund Freud, *Standard Edition, Vol. IV, Preface to the Second Edition of The Interpretation of Dreams*, p. XXVI.) Osama bin Laden's loss of his father also was poignant and seemed filled with ambivalence.

Shame and Humiliation as "Son of the Slave"

Another source of narcissistic wounds in Osama's childhood is found in the way he was treated by his half-siblings in the household. Osama's banished birth-mother, Hamida, was spitefully referred to as Al Abeda (the slave). Osama was cruelly labeled Iban Al Abeda (son of the slave). This constant teasing and devaluation of Osama by his half siblings hurt him deeply and festered in his heart. (Robinson, page 39.)

After Osama's father died, Osama was sent to Tabuk to join his mother in their exile from the rest of the family. He tried to get reaquainted with her. But, Robinson says, "Some of the [bin Laden] family today explain that Osama came to resent both his father for removing him from his mother, and his mother for not attempting to bridge the gap with his father for his sake. The wounds healed but the scars remained." (Robinson, page 40.)

Early Loss of a Best Friend

During his preadolescent and early adolescent years of what must have been inner bitterness, Osama had one true best friend. This friendship with Abdul Aziz Fahd began during the annual father-son desert camping trips with their fathers. They enjoyed periodic visits because their fathers were close, and they also spoke on the phone frequently. But when Mohammad bin Laden died in a helicopter crash when Osama was 10, the two families' relationship changed. The bin Laden family remained respected, but the closeness between the two families ceased. (Robinson, page 59.)

Shortly after Osama returned to Jeddah from Tabuk, he was delighted to learn that Prince Fahd was soon coming to Jeddah for a visit. His best friend Abdul Aziz would be with the Prince. During Abdul's visit, Osama tried dozens of times to contact his friend by phone but was rebuffed. Osama was even turned away from the door of the house where Abdul was staying. Osama has never heard from or seen his friend again. This probably added another element to the acted-out hatred towards the "Father-brother-land" of Saudi Arabia (Robinson, page 60.) In 1994, his Saudi citizenship was revoked because of Osama's increasing threats against the Saudi leaders and America.

Osama's personal adolescent searches mirrored some of the turmoil of the Arab Middle East in the late 1970s. Some authors (Robinson, Bodansky, Dennis) describe Osama as going through typical adolescent rebellion. They say that he was a playboy, drinker, and womanizer in Beirut. Peter Bergen, in his book *Holy War Inc.,* insists that Osama was always serious and devout.

Other than providing quality schools for his son, Mohammad bin Laden let money be his payment to the Piper of parenthood. After his father died, Osama went to high school in Jeddah. He then studied management and economics at King Abdul Aziz University in Jeddah.

Osama's Search for Other Father Figures

Peter Bergen, in his book *Holy War Inc.,* documents the notion that on the surface, bin Laden idealized his father. Osama said that his father was eager that one of his sons would fight against the enemies of Islam, and it is clear that Osama saw himself as that chosen son. Osama told a Pakistani journalist, "My father is very keen that one of his sons should fight against the enemies of Islam. So I am the one son who is acting according to the wishes of his father." (Bergen, page 55.)

Bergen adds, "Having lost his deeply religious [but also very materialistic] father while he was still a child, bin Laden would, throughout his life, be influenced by religiously radical older men."

139

As a psychoanalyst, I find this observation to be important and compatible with my psychodynamic hypothesis about Osama's lifelong ambivalent and hungry search for father figures. I will introduce you to several of these key men.

During his college years, Osama began to show intense interest in Islamic studies. He was greatly influenced by several men. The first, Sheik Yussuf Abdallah Azzam, was born in a village near Jenin, Palestine, in 1941. He was 16 years older than Osama. He graduated with a degree in theology from Damascus University in 1966. Azzam hated Israel, which he blamed for taking Palestinian land. Azzam fought against Israel in the 1967 war. He got a master's degree and doctorate in Islamic jurisprudence at al-Azhar University in Cairo by 1973. He was extremely charismatic. The eloquent Azzam established the worldwide network of *jihad* that won the Afghan war against the Russian communists. Osama and Azzam believed fanatically in the need for *Khalifa,* the dream that Muslims around the world could be united under one devout Islamist ruler.

Azzam was a friend of the famous *jihad* ideologue Sayyid Qtub and the Egyptian sheik Omar Abdel Rahmann (he inspired the 1993 bombing of the World Trade Towers). All three are heroes and mentors of bin Laden and leaders in the formation of a network of holy warriors in the 1980s.

Muhammad Qtub was another of Osama's influential teachers and the brother of Sayyid Qtub (1906-1966), the philosopher and hero of all the groups that eventually joined the Al Qaeda network. Sayyid Qtub was executed in 1966, but his writings in prison have led him to be called the Karl Marx of Islamist global *jihad.* Qtub wrote *Milestones* as well as his masterwork, *In the Shade of the Quran.* Qtub's work has inspired Al Qaeda, Egyptian Islamic Jihad, the Islamic Group (Egypt) and the Muslim Brotherhood (Egypt's fundamentalist movement in the 1950s and '60s).

When Qtub's work is finally fully translated, it will make 15 thick volumes in English. His ideas are articulate, sophisticated, and powerful. He had a traditional Muslim education and had memorized the Koran by age 10. He went to college in Cairo and did further studies in literature. He wrote novels, poems, and a book that is still well regarded, called *Literary Criticism: Its Principles and Methodology*. Qtub traveled to the United States in the 1940s and got a master's degree in education at Colorado State College of Education.

Qtub is so important because his radical Muslim youth-inspiring prose comes not from an academic ivory tower, but from prison, where anti-Western hatreds find their spiritual headwaters. Paul Berman summarizes Qtub's core sophisticated theology of hatred toward the West: "Qtub wrote that, all over the world, humans had reached a moment of unbearable crisis. The human race had lost touch with human nature. Man's inspiration, intelligence, and morality were degenerating. Sexual relations were deteriorating 'to a level lower than the beasts.' Man was miserable, anxious and skeptical, sinking into insanity and crime. People were turning, in their unhappiness, to drugs, alcohol and existentialism. Qtub admired economic productivity and scientific knowledge. But he did not think that wealth and science were rescuing the human race. He figured that, on the contrary, the richest countries were the unhappiest of all. And what was the cause of this unhappiness? — this wretched split between man's truest nature and modern life!" *(The New York Times,* March 23, 2003, page 4.)

This depth of thought and idealism appeals to youth in all societies because of its iconoclastic truth. (I also see a profound conservatism here, nostalgia for an earlier, simpler time that resonates in Western philosophy from Rousseau to Thoreau to many 20th-century poets, songwriters, and others who reject the materialism of modern life and yearn for a pre-industrial Eden of some kind.) But Qtub's writing grew angrier and angrier, and began encouraging martyrdom and violence. It is similar to the way that Abbie Hoffman indirectly inspired violence against the "Establishment" in the 1960s, or, more recently, the way anti-abortion preachers have encouraged idealistic youth to assassinate physicians at abortion clinics in the U.S.

Osama's Life Journey

In 1975, Saudi Arabia's King Faisal was assassinated by his deranged nephew, Prince Faisal ibn Musaid. Osama's Islamist professors/father-figures said the assassin had been driven insane by exposure to Western ways. Similarly, Osama's Islamist Saudi teachers taught that the terrible pain of the Lebanese in their 1975 war was a punishment from God for their sinful acceptance of modern Western materialism and excess. The fatherless Osama was deeply impressed by his conservative Islamist professors who taught that only return to strict Islamism could protect Muslims from the sins and materialism of the "Satanic West." (Bodansky, page 3.)

Another outspoken Muslim and father-figure to bin Laden is Hassan Abdullah al-Turabi, born in 1932 in Kassala, eastern Sudan. Turabi received a secular education in English schools in Sudan. He got a law degree at Gordon University in Khartoum in 1955. He was a clandestine member of the Egyptian Muslim brotherhood even as he earned an M.A. in law on scholarship to the University of London in 1957. He completed his doctorate at the Sorbonne in 1964. Turabi became the spiritual leader of the National Islamic Front (NIF). He became Sudan's spiritual leader under General Omar al-Bashin after a coup in June of 1989. I think his radical Islamist influence is certainly present in the violence and genocide in Sudan today.

In 1989, bin Laden returned to Saudi Arabia, his homeland, after serving as a hero of great renown in the successful *jihad* and defeat of the Soviet Union in Afghanistan. It appeared that Osama would settle down, but on August 2, 1990, Iraq invaded Kuwait. Osama presented a detailed plan for self-defense of the Saudi kingdom, bolstered by battle-hardened Saudis who had fought in Afghanistan. Bin Laden offered to recruit the "Afghans" and said his family construction company could quickly build fortifications. He warned that inviting or permitting infidel Western forces into the sacred lands of the kingdom would be against the teachings of Islam and would offend the sensibilities of all Muslims.

Osama was not alone in these concerns among the Saudi leadership, but his offer was ignored despite his great popularity at the time. When he grew more critical and insistent, the Saudi leadership threatened his extended family. In 1991, fearing reprisals for his outspoken anti-American and anti-Saudi pronouncements, bin Laden bitterly left for exile in the land of a more friendly father-figure: Turabi of the Sudan. Turabi became Osama's mentor/father-figure and Islamist colleague.

Osama, I believe, was perpetually (unconsciously) caught between his new Islamist beliefs and the dazzling financial success from his dead father's affiliation with the materialistic Saudi leaders and their American "friends." Today it seems to me that Osama seeks both the destruction of the father-figures of the Saudi regime that disowned him, and the casting out or destruction of the Saudis's American "friends." On a conscious level, he idealized his father, but on an unconscious level he held tremendous ambivalence toward even his admired mentor Azzam.

My psychoanalytic hunch is that unresolved grief, disappointment, and shame over his dead father's Western materialism seem to have focused and magnified Osama's identity as a self-appointed anti-Western radical Muslim warrior chieftain. Osama's relentless, grief-laden search led him to find other radical, admired and admiring father-figures. His relationships with Azzam of Afghanistan and Turabi of Sudan filled the bill. Bodansky, in his book, *Bin Laden: The Man Who Declared War On America,* documents in great depth and detail the mutual admiration reverberating between Osama and these men.

Ayman al-Zawahiri, the Egyptian physician turned radical Islamist, and Mullah Muhammad Omar, leader of Afghanistan's Taliban, became Osama's brothers in rebellion and rage. They also served as a replacement for Osama's Saudi half-brothers, who eventually disowned him over his terrorist commitments.

Enter Osama's Idealized Older Brother Surrogate
Dr. Ayman al-Zawahiri was born in Egypt in 1951, six years before Osama bin Laden's birth. He comes from an aristocratic, affluent Egyptian family. He was educated as a pediatrician. In 1973, as a

medical student, he joined with an electrical engineer and an army officer to form the Egyptian Islamic Group, dedicated to the violent overthrow of the Egyptian state. He served as a medical officer in Afghanistan, where he was regarded as a hero, along with bin Laden, in the *jihad* against the Soviet Union. He was very close to Egyptian sheik Omar Abdel Rahman, the so-called blind sheik behind the 1993 bombing of the World Trade Towers and the assassination of Egyptian President Anwar Sadat in 1981.

Zawahiri has grown very close to Osama (as an older brother/ father figure) over the years of their alliance in building what Bergen calls "Holy War Inc." Many experts feel that al-Zawahiri is the man who influenced Osama to morph from a donor of money to a warrior of violent *jihad*. He is a constant companion and mentor of bin Laden. In my opinion, Zawahiri has psychologically replaced and compensated for Osama's painful loss of his best friend, Abdul Aziz Fahd, in his childhood.

Osama was a heroic, fearless military leader in Afghanistan's successful war against the Soviet Union. For bin Laden, the Afghanistan war was a profoundly spiritual experience. (Bergen, page 61.) This enabled him to be both an admired warrior son of radical terrorist state leaders, and a legendary hero/older brother/ mentor for millions of angry young Muslims. We can see how money has never been able to compensate for Osama's grief over and ambivalence about a father who was not around when the adolescent Osama needed his guidance most. Fathers of adolescent sons often feel superfluous or devalued by their sons. But, beneath this surface of rebellion in the service of fledgling independence and identity searching is a desperate need for a father with moral strength and dignity! All Malignant Pied Pipers seem to have unconscious preoccupations with ambivalent longing for, but also hatred of, the father who did not stand up to them and for them! This leads them to be the ultimate father/themselves unto themselves. I believe Osama is a soul brother to other Malignant Pied Pipers.

Rohan Gunaratna, in his book *Inside Al Qaeda: Global Network of Terror,* gives good examples of Osama's severely ambivalent conflicts with even his mentor/father figure Azzam. Azzam felt that noncombatant women and children should not be killed in the *jihad.* In his strict Wahhabi fundamentalism, Osama felt that all infidels should die — even women and children. Gunaratna even raises the question of whether Osama held symbolic patricidal urges, as he may have participated in Azzam's assassination! (Gunaratna, pages 30-33.)

On April 7, 2002, *The New York Times* published a poem, "The Travail of a Child Who Has Left the Land of the Holy Shrines," co-authored by Osama bin Laden and the poet Dr. Abd-ar-Rahman al-Ashmawi. Half of the 42 verses were written by bin Laden. Eighteen verses mention Father, Son, Brother, Child, or Family. One line mentions Mother. I record these selected 18 verses (with my own commentary) because of their reflection of bin Laden's search for his father/himself.

Father, where is the way out? *[of all our troubles?]*
when are we to have a settled home?
Oh, Father, do you not
See encircling danger?
Long have you made me travel, father,
Through deserts and through settled lands.
Long have you made me travel, father,
In many a sloping valley,
So I forget my kinsfolk,
My cousins, and all men.
Why has my mother not returned? *[Early childhood separation?]*
How strange! Has she taken a taste for travel?
You, father, do not crave
an easy living from mankind. *[Actually he did!]*
Tell me, father, for I find
No brief enlightening explanation.
'Forgive me, son, for I am stuck
Both powerless and speechless.'
Forgive me, son, for nothing do I see

In our terrain but a declivity.
As, marked by manliness and pride,
Does our commander, Mullah Omar. *[Leader of the Taliban and clearly a father / older brother figure; he also married one of Osama's daughters]*
Why, father, have they sent
these missiles thick as rain,
Showing mercy neither to a child
Nor to a man shattered by old age?
Father, what has happened
so we are pursued by perils?
It is a world of criminality, my son
Where children are, like cattle, slaughtered.
Zion is murdering my brothers,
And the Arabs hold a congress!
Why have they not equipped a force
To shield the little one from harm?

Bin Laden's poem lines can be seen as free associations similar to those a psychoanalyst hears from the couch. In other lines of this poignant poem, bin Laden reveals his search for a home:

A decade has passed, whose years were spent
In homelessness and wanderings.
I have migrated westward
To a land where flows the Nile. *[Clear reference to father surrogate Turabi of Sudan]*
Of Khartoum I love the character,
But I was not permitted to reside.
So then I traveled eastward
Where there are men of radiant brows.
Kabul holds its head up high *[Personifications of idealized fathers]*
Despite the hardship and the danger.
Kabul, with a shining face,
Offers all-comers shelter and help.

Often when captured by the poetic muse, the poet reveals his or her inner feelings about the core of self. Poetry of this sort provides access to the more unconscious levels of the poet's psyche.

In the terrorist training camps of Afghanistan, bin Laden both found and became his own lost father, brother, mother, and family. Osama bin Laden's psychological father-hunger has been relentlessly and destructively acted-out with an uncanny resonance with thousands of father-hungry young Muslims around the world. Bin Laden's apparent massive denial of his own grief, loss, and fear of being alone, has become part of the destructive illusions and delusions he offers his young terror cult followers. He cloaks these illusions in the radical Koranic interpretations offered by his mentors. Muslim clerics who are sympathetic to bin Laden's radical Islamist theology of hatred teach in *Madrassahs* all over the Muslim world. They act as lieutenants to recruit and indoctrinate followers for Al Qaeda and its warrior prophet bin Laden. Osama and his colleagues from Al Qaeda have even seduced American youths John Walker Lindh and Jose Padilla with their deadly music. Osama bin Laden has become the ultimate Malignant Pied Piper, and his music goes on and on ...

Paying the Piper

My study of destructive cults has revealed remarkably similar Pied Piper behavior among leaders and followers, from Jim Jones to David Koresh and now to Osama bin Laden's Al Qaeda terror cult network. In this model, charismatic leaders with sadly distorted personalities lead idealistic, father-hungry, or disillusioned young people away from their home villages. In Osama's and other Al Qaeda gurus' cases, they not only seduce the children, they also seduce the whole village!

The price all competent and loving parents pay (or don't pay) the Piper of parenthood, is quality time. This means significant chunks of time devoted not to what the child can do to promote the self-esteem, reputation, or psychological well-being of the parent; but towards what the parent can try to do to help the development of solid identity, self-esteem, realistic self-confidence, and solid moral development of the child. This investment of parental time is especially important during a particularly intelligent and perceptive child's adolescent identity crisis.

When his father died, Osama was probably left feeling bitter, humiliated, alone, and, eventually, narcissistically enraged at an unconscious level. The Rev. Jim Jones got his Piper-payback in the form of 913 lives at Jonestown in 1978. Osama bin Laden got thousands of lives as Piper-payback on 9/11 alone. Osama's radical Islamist Pied Piper music has led many young Muslims away from moderate Muslim villages where parents could never pay the Piper enough.

Osama bin Laden's Dark Epiphany
The early childhood disappointments, shame, and empathic parental failures of future destructive cult leaders are further magnified and compounded by dark epiphany experiences in late adolescence or young adulthood. This is true for Osama bin Laden. Let's backtrack a little to focus on one of Osama's dark epiphanies.

Osama returned to Saudi Arabia in 1989 as a hero of the war against the Soviet Union in Afghanistan, but events of the next two years left him feeling rejected and threatened by the Saudi leadership, who ignored his offers to help defend the country when Iraq invaded Kuwait. This dark epiphany slowly smoldered into molten rage and vengeance toward the betraying Saudi father-figures and their American "infidel friends."

Our Lost U.S. Opportunity for Friendships
In Afghanistan after the Soviets left, America and our Saudi "friends" did *not* pay the Piper. We abandoned a crumbled Afghanistan, and a Robin Hood-like Pied Piper called bin Laden emerged. Now there are many future bin Ladens born daily around the world. Our foreign policy does have life or death consequences for us! America can no longer give billions to repressive dictators in Arab countries with no benefits to the ordinary people.

A Confounding Dynamic: The Uncanny Impact of TV Media
Cults and terrorism as a means of social rebellion and control have existed since antiquity. In a confounding, paradoxical way, modern worldwide instantaneous media coverage provides a perfectly magnified stage for the terrorist's ghastly reality drama. If terrorists had to buy television time to cover their events, they would have to pay millions of dollars for what they get for free. The media provides both publicity for the terror event and grandiose narcissistic affirmation of the terror cult and its leader. These issues are clearly in evidence when one studies Osama's hubristic videotape performances. The terror cult leader becomes an exciting, flamboyant, and darkly charismatic television celebrity.

Many people who saw the World Trade Towers hit by airliners said, "It was surreal!" or "It looked like a Hollywood movie." In fact, good moviemakers or horror fiction writers exploit the same experience of "the uncanny" that a terrorist does. (Sigmund Freud (1919), "The Uncanny." This lesser-known paper by Freud should be read by every fiction writer. Stephen King breathes it in his work.) The intuitive and/or calculated impact of the terrorist act comes from our tendency to regress to magical, black-and-white, superstitious, or primitive thinking under stress or shock. Our shocked nervous systems and minds become more fragile and vulnerable. Osama bin Laden even had issued threats and warnings months before his terror cult's actions. This gave a magical, grandiose, prophetic quality to the 9/11 events.

Formation of the Self
Psychoanalytic theorists such as Kohut locate the core of self-formation in the second year of life. (Heinz Kohut (1971), *The Analysis of the Self,* pages 64-67.) Kohut moved beyond classic Freudian theory about self-formation and explored the importance of empathy in parent-child relationships. The young child seems to have two basic experiences of self-love. At times there is a sense of grandiose-exhibitionistic self-delight; and at other moments, an admiration and resonance with the idealized image of the parent or their surrogate childhood hero.

The smooth, steady, and healthy integration of the self is facilitated by responsive, empathic, and confidently affirming participation by parents or their surrogates. Empathic parenting helps modify the child's early grandiose fantasies of undue power, ambition, and significance in interaction with the strength of parental heroes. This socialization process provides realizable goals, values, and guiding principles for the young individual. These early, crucial, transitional self-other experiences with parents or role models provide models in the young mind for self-love, self-esteem, and confident, responsible actions.

Group Self, a Crucial Idea

There is a parallel process by which an individual's sense of himself as part of a group is formed. (Kohut, *The Search for the Self,* pages 837-838.) In essence, inner representations of our Self and our Self-in-a-group are parallel and conjoined early developmental and maturational experiences.

Vamik Volkan has observed that children often experience angry or destructive feelings as dangerous or alien. These dangerous impulses are often displaced on to the external world of "others." (Vamik Volkan, *The Need to have Enemies and Allies: From Clinical Practice to International Relationships,* pages 27-34.) This externalizing process is the way we all form a concept of "The Enemy," as really a projected part of our own feelings. As our sense of Self and Self-in-a-Group is forming, there are other influences besides the experiences of our parents. These have been called "Extensions of the Self," or extensions of the Group-Self experience. These Self-extensions are the national geography or terrain; the flag, monuments, churches, temples, mosques, or synagogues. They also include charismatic political figures from outside the family. The cohesive self formed in preadolescent years is vital for subsequent, secure identity formation in one's cultural or community group. In times of societal stress, deprivation, persecution, or crisis, charismatic leaders like Osama bin Laden serve as messianic extensions of the group self or identity formation. It is at this location that youths are vulnerable to Islamist Malignant Pied Pipers.

Most psychologists agree with Erik Erikson, who thought that individual identity is consolidated with enthusiastic intensity in the adolescent years. Charismatic/messianic culture heroes and anti-heroes have particularly important impact on self and identity formation in young people in countries where parents and families are forced into humiliating refugee camps or other painful group exiles. In his 1986 book, *After the Last Sky: Palestinian Lives*, Edward Said presented poignant examples of families in these dilemmas. This vivid quote illustrates the humiliation of the refugee experience (Said, page 14):

"When A.Z.'s father was dying, he called his children, one of whom is married to my sister, into his room for a last family gathering. A frail, very old man from Haifa, he had spent his last 34 years in Beirut in a state of agitated disbelief at the loss of his house and property. Now he murmured to his children the final faltering words of a penniless, helpless patriarch. 'Hold on to the keys and the deed,' he told them, pointing to a battered suitcase near his bed, a repository of the family estate salvaged from Palestine when Haifa's Arabs were expelled. These intimate mementos of a past irrevocably lost circulate among us, like the genealogies and fables of a wandering singer of tales. Photographs, dresses, objects severed from their original locale, the rituals of speech and custom: much reproduced, enlarged, thematized, embroidered, and passed around, they are strands in the web of affiliation we Palestinians use to tie ourselves to our identity and to each other."

Let's consider Peter Bergen's reminder that bin Laden's hero-mentor, Azzam, was born in a small village near Jenin, Palestine, in 1941. The hatred of Israelis and radical belief in *jihad* simmered in Azzam and his colleague at al-Azhar University in Cairo, Sheik Omar Abdel Rahman (the blind cleric). These hero mentors and their hero worshippers, Ayman al-Zawahiri (older brother-figure) and bin Laden, make clear the symbolic emotional connection for bin Laden between Egyptian *jihad,* the Palestinian cause, and his search for a radical Islamist "father" to please. The radical Islamist *jihadists* want to bury us!

Osama's childhood losses and the generic Palestinian pain situation described by Said have a profoundly important impact on youth. Zawahiri and bin Laden found intense bonds of friendship. They both had subtle but important disappointment in and rebellion toward their aristocratic and wealthy families of origin. They formed a new family of Al Qaeda.

The Profound Dangers of Groupthink

In the global radical Islamist *jihad* leadership brain trust (Osama bin Laden, Azzam, Zawahiri, Turabi, Rathman, et al.) I believe we can see elements of groupthink. (Irving Janus and Mann, I., *Victims of Groupthink: A Psychological Study of Foreign Policy Fiascoes.)*

Janus and Mann explain the Bay of Pigs mistake by John Kennedy and his entourage according to the "groupthink hypothesis." Janus wrote, "Members of any small cohesive group tend to maintain esprit de corps by unconsciously developing a number of shared illusions and related norms that interfere with critical thinking and reality testing."

The major symptoms of groupthink Janus cited are:

1. A shared illusion of invulnerability, which leads to excessive risks.

2. A shared illusion of unanimity based on the reluctance of individuals to voice their doubts, and the assumption that silence implied consent.

3. The related tendency of group members to self-censor their own misgivings and in some cases to discourage others from sharing their concerns with the leader.

4. A tendency for the leader to handle the group's meetings in a way that discouraged discussions of the drawbacks of the emerging plan.

This common group phenomenon can occur in the Bush Cabinet discussing preemptive war or in Osama's *jihad* conclaves!

My hunch is that such groupthink occurred at the many Turabi-bin Laden-Zawahiri planning meetings of Turabi's Islamic Charter Front during 1991-1993. (Bodansky, pages 32-33.) I think that pervasive group-death-think occurred as the worldwide *jihad* plans unfolded. Azzam tried to break through the Al Qaeda "group jihad deaththink."

Gunaratna says that when Osama's beloved mentor Azzam opposed the slaying of women, children, and noncombantants in the *jihad* terror plans, Osama turned on his father figure and probably colluded in his assassination!

How Terrorist Leaders Become Heroes to Psychologically Wounded or Disaffected Young People

Al Qaeda finds foot soldier recruits in Muslim community situations where parents are humiliated, devastated, and even killed in plain view of their children. Poverty-stricken Muslim refugee communities become spawning grounds for future terrorists. The narcissistic injury to the parent results in wounds to a child's inner hero. This of course does not always lead to a career as a terrorist, but vulnerable youths turn outward towards the Extended Self for heroes. Enter the enchanting music and mystical charisma of a rebel hero like Osama bin Laden. He spuriously offers to lead the injured individual and group-self out of psychological exile by turning passive pain into bold actions.

Why Wounded Sheep Find Familiar Pied Piper Music Profound

Bin Laden is himself an outcast and disowned exile from his country and his large family of Saudi millionaires. I hypothesize that he has tried to heal his own disappointments, narcissistic injury, and rage by becoming the self-appointed savior, martyr, and older brother/father/messiah for his followers and admirers. He has been largely successful as is evidenced by the cheering Muslim crowds in many places in the world after 9/11.

Osama bin Laden's role as a terrorist leader allows him to act-out his unconscious inner narcissistic rage at his father, mother, siblings, rejecting homeland, and Saudi Arabia's oil customer/ally/friend America. In this sense, he becomes like most destructive cult leaders. Their deepest motives have to do with power, control, revenge, and overcoming a desperate fear of aloneness and meaninglessness. They gain a sense of power and mastery over their own childhood psychological deformities and feelings of insignificance by becoming overwhelmingly significant and powerful in the lives and destinies of their followers. These Malignant Pied Pipers enter a relentless quest to become "strong" parental figures for other people in order to finally experience a "good" parent within themselves. In Osama's case, he uses the radical, distorted interpretation of the Koran by his mentor/father-figures to mask his inner psychological motivation.

This long search requires a ready supply of child-admirers, whom they find in their cult recruits. Beneath the outward confidence, smoothness, and swagger of Malignant Pied Pipers is an unconscious sense of shame, rage, and fear of humiliation or aloneness. The terror cult honeymoon is eventually over and the formerly neglected and abandoned leader now becomes the neglecter and abandoner. Evidence for this occurred in bin Laden's videotape issued after the 9/11 events. He gloated and smugly chuckled about his clever strategy of not informing most of the terrorist plane hijackers about the essence of their suicide death missions.

Apocalyptic Scenarios: Group-Self Death and Rebellious Martyrdom in Terror Cults

The terrorist cult leader becomes a legend in his own mind and a pseudo-healing legend for the wounded group-self of those who feel rescued. Rebellious charisma in a terror leader meshes with the followers' narcissistic passive-receptiveness to his charismatic influence. The leader-follower pattern in terror cults is remarkably similar to what is seen in apocalyptic cults like those of Jones, Koresh, and Asahara.

The group death or martyr scenario gives the terror-cult group a special, exciting, and dramatically triumphant defining martyr myth. It becomes a source of "underdog" heroism, and paradoxical group cohesion and identity. The individual and group will triumph over Evil. These group death myths represent the ultimate *denial of death* as described by Ernest Becker in his 1973 Pulitzer Prize-winning book, *The Denial of Death*.

For bin Laden, the motivating apocalyptic scenario is his assertion that all Muslims in the world are being threatened by the West, particularly by Americans and Jews. In a book bin Laden wrote in 1998, he called the faithful to a global *jihad*, a "new vision" that demands the deaths of all Americans and Jews, including children. To attain this cynical and religion-perverting vision, any violence is justified, from terrorist bombings to suicide missions.

Evil is defined initially by the leader bin Laden via "Fatwa," but gradually becomes co-authored within the group-self as their group salvation death myth. The codependent leader holds the martyr death myth out to the followers as magical reward. The terror cult leader also holds the death myth over the heads of the followers so as to magnify the special domain of his "mana" power and self-importance. The leader is needed for the dramatic destructive action that is being planned, for which no one individual takes personal responsibility. The leader experiences the ultimate "celebrity" and fantasized triumph over his life-long insecurity, hurts, and fear of aloneness.

The codependent followers embrace the martyr death idea because it brings heightened meaning to their otherwise humdrum or miserable lives. I have listened to hundreds of hours of recordings of Jim Jones's ranting free-associative "white night" sermons in Guyana. I was fascinated by the comments of his followers in the background. Those comments extol and reveal the group idealism, excitement, and grandiosity of the group-death scenario. Group death myths for such grandiose causes are truly terrifying. The inspirational and instruction manuals of the 9/11 terrorist pilots reflect these ultimate denials of death and devotion to the pathetically alleged cause of "Allah," as inserted by bin Laden!

The Children of Terror and Terrorism

Psychologist Rona Fields studied adult terrorists and children exposed constantly to terror in their environment. Some of Fields's interviews and psychological testing included children who years later were retested after they had become terrorists as adults.

Fields concludes that children who observed frequent violence in Northern Ireland, Palestinian refugee camps, and black South African townships have a sense of helplessness not changed by anything their parents can do. The experiences lead these children to feel stuck at a stage of moral development where they view right and wrong exclusively in terms of their group-self and black/white thinking. Fields observed that Protestant and Catholic youths in Northern Ireland have identical psychodynamic and moral development. They are obsessed about right and wrong, yet they have low anxiety about their anger and high curiosity about violent solutions. As these children of violence seek novel and adventurous outlets for their anger, we can see the fertile soil for the conversion of the terrorized into the terrorist(s).

Diagnosis

Osama bin Laden exhibits eight out of nine criteria for Narcissistic Personality Disorder:

1. Grandiosity and self-importance: bin Laden broadcasts his Pied Piper music via Arab and international television networks in his self-appointed role as wealthy rescuer of oppressed Arabs.

2. Fantasies of success, power, and brilliance: Bodansky describes bin Laden waving his copy of the Koran and quotes bin Laden saying, "You cannot defeat heretics with this book alone, you have to show them the fist!" Bodansky goes on about bin Laden, "He also elucidated his vision of the relentless, fateful, global *jihad* against the United States in a book titled *America and the Third World War.*

In this book bin Laden propounds a new vision, stressing the imperative of a global uprising. Bin Laden in essence calls on the entire Muslim world to rise up against the existing world order to fight for their rights to live as Muslims — rights he states are being trampled by the West's intentional spreading of Westernization." (Bodansky, page 388.) No compromises – it is Osama's way or no way. Is there anything more omnipotent and grandiose (and chilling) than Osama's legend in his own mind — and so many Muslims' resonance with it?

Bin Laden has associated with fugitive Taliban leader Mullah Omar, to whom he married his daughter. Osama's fourth wife is either Omar's daughter or a close tribal relative. (Bodansky, page 307.) Osama has associated with Ayman al-Zawahiri, the head Egyptian terrorist and his terror commanders. (Bodansky, page 115.) Along the road to grandiose splendor, Osama chose father figures like Turabi of Sudan and Azzam, and learned sly tactics as their colleague and acolyte.

3. Specialness: To my observation, bin Laden clearly enjoys the TV spotlight during his performances.

4. Requires excessive admiration: This seems apparent in his dealings with the Saudi leaders, when he proposed to form an army to defend Saudi Arabia and was narcissistically wounded at the Saudis' rejection of his plan. Rather than admire him, the Saudi leaders ignored his offer.

5. Entitlement: see #4. Bin Laden expected instant compliance and admiration.

6. Exploitive of others for his own ends: In my opinion, bin Laden exploited some of the 9/11 skyjackers because he

laughed when it was announced that some of the "martyrs" were unaware of the full suicide plans for the hijacked planes. He certainly doesn't send his own son on such a mission.

7. Lacks empathy: bin Laden shows no empathy for the victims of his bomb attacks or 9/11. Even innocents! Even his own countrymen.

8. Envious: Osama seems too grandiose and malignantly narcissistic to envy anyone.

9. Arrogant: Practically every TV broadcast and public statement by bin Laden reeks of arrogance, grandiosity, and fundamentalist mentality.

Kernberg's Criteria for Malignant Narcissism:

1. In the context of ubiquitous Arab anti-American feelings, I do not see Osama as clinically paranoid or psychotic.

2. Osama is chronically homicidal in the context of his bogus *jihad* of hate.

3. I think bin Laden is psychopathically manipulative and dishonest with his followers and with the Arab world.

4. At conscious and unconscious levels I think bin Laden showed clear evidence of "joyful cruelty" as he watched the suffering he helped create on 9/11.

The psychodynamic configurations of terror cult leaders like Osama bin Laden are remarkably similar to apocalyptic/destructive cult leaders. The group dynamics and powerful *denial of death* motivations are also similar, but vastly wider in scope, significance, and impact. Malignant Pied Pipers live on. They multiply like mushrooms in a dark forest. They are far more powerful than nuclear mushroom clouds.

The future generations of Osama bin Laden's terror cult will not be eliminated by bullets, missiles, or "smart-bombs." The future foreign policies of America need to be informed about the social-self psychology of the terrorist group and its leaders. Devastated communities, families, and wounded group-selves in war-torn nations continents away are ignored at our peril. Billions of military aid dollars given to Middle East "friends" may not be worth the enemy-accumulating consequences. "Nation-building" may be impossible, but we cannot just walk away. Genuine foreign aid helps wounded world communities find ways to rebuild their dignity and group-self confidence — not terrify them into submission, just so we will feel safe.

Fundamentalism in Our World

Fundamentalists, no matter how sweet, kind, or pious they seem on the surface, are convinced that they have *the* superior moral, ethical, theological, epistemological, political, and spiritual truth. Fundamentalists (Christian, Jewish, Muslim, Sikh, Hindu, even some anti-war pacifists) experience in a delusional way their own way of thinking and believing as the one and only way. Fundamentalist mentality is more prevalent and seems more readily embraced at times of severe social turmoil, rapid social change, economic hardship and oppression of minority groups.

The radical fundamentalists of any religious or political group are willing to confidently condemn to hell and/or kill those who do not believe exactly as they do. They are the final judges, juries, and "holy" executioners. There is no room for honest discussion, doubt, or debate. They are their own gods themselves.

Our new century seems to be plunging relentlessly and exponentially into worsening pathology. Misunderstandings between East and West, moderate Muslim and radical Muslim are growing. Destructive, polarizing fundamentalist mentality pollutes the group self and seems to be leading deeper into the maelstrom of *jihad,* preemptive war, and the so-called "war on terror." (A war against a tactic.) Fundamentalist groupthink seems to afflict both the leaders of *jihad* against America, and our American leaders in the "war on terror" and their groupthinking entourages.

Without creative, empathic leaders, group-war-think can spread to western electorates and the mushrooming followers of worldwide radical *jihadists.* Sadly, it seems possible for the world's "collective unconscious" to be dominated by fear, distrust, and group-war-think.

CHAPTER EIGHT

WHY WE STUDY THE MINDS OF MALIGNANT PIED PIPERS & THEIR FOLLOWERS

My main purpose in writing this book is to provide a depth of psychological information about destructive and exploitive cult leaders and their followers, so that mental health professionals, clergy, teachers, law enforcement professionals, and all educated persons can be aware of the dangers these cults embody. As a way to summarize some key issues, I wanted to discuss some general conclusions, cautions, and recommendations about MPPs. Then I will examine specific dangers and dynamics posed by each Malignant Pied Piper I have discussed. I hope that close examination of each MPP will help people to spot these mesmerizers quickly, and find effective action plans in dealing with their deadly siren songs.

General Observations
A careful reader of this book might ask why many people who have suffered childhood neglect, abandonment, and painful narcissistic or shame-filled humiliation experiences with parents *don't* become destructive cult leaders or followers. In other words, the childhood histories (deformities or molding experiences, discussed in Chapter One) and even the magnifying and compounding dark epiphany experiences of young adult life that I describe in MPPs, are *necessary* but not *sufficient* to explain all domains of causality behind these destructive cult leaders. In fact, it is fascinating to take

note of profoundly benevolent and altruistic leaders who have experienced narcissistic injuries in childhood, but managed to transcend these difficulties to turn lemons into lemonade with their lives. (See, for example, *The Immortal Ataturk* by Vamik Volkan and Norman Itzkovitz, and *Gandhi's Truth* by Erik Erikson.) However, I think the psycho-biographical profiles and descriptions I provide about the modus operandi of these MPPs can be useful as preventative medicine. We need constant reminders about destructive cults, how they develop, our denial or vulnerability to them, and awareness of who their agents are.

People tend to protect themselves by starkly demonizing MPPs like Hitler, bin Laden, Manson, Jones, Koresh, Asahara, Applewhite, Jouret, DiMambro and their followers, as if they were some alien species. In fact, even the severely mentally ill or serial killers are not psychotic or murderous 24 hours a day. MPPs can be very winsome and appealing at early encounters. It can be valuable to be able to spot subtle patterns of a potential MPP or the recruitment techniques that they or their cult group use.

All people to varying degrees have a basic need to be a part of a community. It is healthy and essential to maturation and personality development to affiliate with and have meaningful experiences with small and large groups. Groups provide healthy avenues to create meaning in our lives. Patriotism and religious loyalty, for example, are normal. Even atheistic faith can be respectful and dignified. Thankfulness and respect for the wonderfulness of America is not incompatible with vigorous criticism of its policies or leaders. MPPs do not tolerate vigorous criticism or intelligent questioning of their teachings for very long. They may feign tolerance for a while, but down deep, they think they know all the answers. It is their way or the highway. This is a key area to be mindful and watchful about with leaders of groups. MPPs can exploit and manipulate our inherent need to affiliate with groups. I have shown this in various vignettes in this book.

Beware of Any Guru
"When I was a child I spoke and thought and reasoned as a child does. But when I became a man [adult], my thoughts grew far beyond those of my childhood, and now I have put away the childish things." I Corinthians 13:11

It is psychologically healthy to retain some of the capacity for wonder and curiosity that abounds in a child. But a psychologically and spiritually healthy adult person should muster a nondefensive but firm skepticism toward any charismatic leader who claims to have most or all the answers to life and how it should be lived. It does not matter if the know-it-all guru is a radical Islamist cleric or a Western fundamentalist preacher. This is especially true if TV or Hollywood star personalities endorse the guru and his or her group.

This process of healthy questioning should be especially vigorous if this charismatic person shows sudden, personal-space-invading interest in you or your family. Challenge the guru's reasons for such personal interest in you. Beware when they give either scanty or self-promoting and showy information about themselves or their motivation. Be actively curious with them about their credentials and those of their group. Do they have a board? A charter? A seminary-training certificate that you can verify? Is the guru a fulltime minister, priest, pastor, rabbi or clergyperson? Is there any formal literature or scripture that you can read and discuss with friends or family before any further discussion will proceed? If he or she grows haughty or defensive about these questions, you have detected the probable presence of an MPP. Do not let them touch you, pray with you, or require any financial commitment from you at such a sudden encounter.

Destructive Cults Are Not Just Innocent or Benign "New Religious Forms"
Some of these groups have clever recruitment techniques, such as "love-bombing," and may appear innovative and intriguing at first. They may seem idealistic or benignly utopian for a while, until the leader begins to show his or her incremental but relentless journey

into grandiosity, malignant narcissistic personality deterioration, or other acted-out personal agendas. Jim Jones, in the early days of his cult's trajectory, set up soup kitchens and housing for the poor. Jones received letters of praise and commendation from first lady Rosalyn Carter and vice president Walter Mondale about his compassion and community service.

Helping a Cult Victim

As we described in the story of Cindy in Chapter Six, once a person is ensconced in a cult it is very difficult for them to disaffiliate. "Deprogrammers" often provide a treatment than is worse than the disease. It is important for friends and family to work tenaciously to keep the continuity of the relationship. Keep your sense of humor and don't give up. Cult leaders and their lieutenants don't make these efforts easy. Family members should seek therapy, advice and support for themselves.

When a family member leaves the cult, give them active love, praise and support. Individual and family therapy is often necessary. Make sure the therapist is well informed about the post-traumatic effects of a cult experience and the impact of cult indoctrination techniques on the mental status of a victim. It takes years to mature away from a cult's domination. However, knowledge and vigilance in these matters may save a life and/or a soul.

Specific Lessons from MPPs

1. The Eyes, Hypnotism, and the "I" of Self.

Followers of Jones, Koresh, Applewhite, and Manson have commented on the uncanny, hypnotic experience they had in encounters with these Malignant Pied Pipers. Deborah Layton, in her poignant and perceptive book *Seductive Poison: A Jonestown Survivor's Story of Life and Death in The People's Temple,* describes her first encounter with Jim Jones at age 17:

"Grabbing hold of my hands, he looked into my eyes. I could feel the heat of his gaze, which burned like white-hot coals. 'Debbie, you have wandered upon this earth looking, wanting, and needing answers. I can give you them. For every unknown in your mind, I can give you enlightenment. For your fear, I can give you strength. For your sorrow, I can give you hope and a dream we will attain together.' I suddenly trusted him completely. I wanted to scream, Yes, I do, I love you."

Applewhite's picture on the cover of *Time* (April 7, 1997) shows the guru's hypnotic intensity of gaze displayed on the Heaven's Gate alluring web site. Manson seems both to have used penetrating eyes on his followers and to enjoy giving the eye to photographers, journalists, prosecutors, and observers in the courtroom. Koresh flashed aggressive eyes during his sermons in Waco. Asahara is legally blind but got many followers to "see" metaphorically, believe his absurd predictions, and even kill for him. Many cult recruiter lieutenants are instructed to stare at a potential recruit with a focus at several inches behind the center of the recruit's forehead.

In his fascinating paper "The Uncanny" (1919), Freud discusses the "Evil Eye" (page 240), the fear of going blind as fear of castration, and the penetrating phallic aspect of staring (pages 231-232). Osama bin Laden's TV presence makes use of the seductive and evil eye. In his paper "Group Psychology and the Analysis of the Ego," Freud cautions us about the face of the hypnotist:

"The hypnotist awakens in the subject a portion of his archaic heritage which has also made him compliant towards his parents and which experienced an individual re-animation in his relation to his father; what is thus awakened is the idea of a paramount and dangerous personality, towards whom only a passive masochistic attitude is possible, to whom one's will has to be surrendered, while to be alone with him, 'to look him in the face,' appears a hazardous enterprise."

It is also a seductive experience, as MPPs demonstrate. In a lonely "in-betweener" or an alienated, angry, or narcissistically injured teenager or young adult, the hypnotic intensity of an MPP is hard to fight off. This is particularly true if peer group attractions accompany the cult leader's seduction. Bugliosi, in *Helter Skelter* (page 653), commented that every member of Manson's "Family" had dropped out and was angry at society before ever meeting Manson.

Michael Langone (*Psychiatric Times,* July 1996, page 14), describes susceptibility to trance-like states as one of ten situational or developmental features that singly or in combination make people more receptive to cult sales pitches. The other nine he describes are: a high level of stress or dissatisfaction; lack of self-confidence; unassertiveness; gullibility; desire to belong to a group; low tolerance for ambiguity; naïve idealism; cultural disillusionment; and frustrated spiritual searching. We need to be aware of these vulnerabilities and help youths not to fear a stranger's eyes.

2. Distorted use and abuse of the Bible, Torah, Koran, or other scripture or literature.

Jim Jones used distortions of the Bible and actual anti-biblical statements pretty early in his preaching. This progressed to flagrant declarations of himself as God or Messiah. Koresh at first impressed peers and followers with his encyclopedic knowledge of the Bible but launched relentlessly into his own self-focused prophecies and grandiose declarations about himself as a prophet-God. Osama bin Laden's *fatwas* and their murderous content spring from his inner projections and acting-out of his own unconscious hatred of parental figures.

In this area we can see the potential value of theologians, philosophers, and all clergy in countering the bogus hyperbole of MPPs. I have been disappointed that more moderate Muslim clerics have not confronted bin Laden's blasphemy or the atrocious suicide bomber behavior of Hamas. Moderate Jewish rabbis need to confront the destructive Jewish fundamentalists in Israel as well. I would suggest that passive masochistic or sadistic narcissism I have discussed regarding follower behavior may be involved in this issue.

3. Control of romantic intimacy and sexuality.

Finally, I want to point out that any group leader that begins to try to control or dictate the boundaries of romance, intimacy, and/or sexuality is showing MPP danger signs. Applewhite's advocacy of castration to control sexual desires is an extreme. Jones and Koresh's personal sexual boundary problems and excesses, while matchmaking and asserting control of their followers' sexuality and marriages, was a red flag early in their reigns.

EPILOGUE

"In-betweeners," Spiritual Hunger and Vulnerability to Seduction by a Malignant Pied Piper

Hitler, Jim Jones, Osama bin Laden, and other Malignant Pied Pipers can be prevented or defeated. Their group evil can be headed off by dedicated remembering, intuitive perceptiveness, and action-oriented, in-depth understanding. This knowledge can alert us to trouble ahead. It is crucial for us to know ourselves and our enemies.

Let's return to the question that haunted me as I decided to write this book.

"How on earth could someone as intelligent, perceptive, and caring as Tim Stoen have been duped by a grandiose conman, an Elmer Gantryesque jerk like Jim Jones?" That question to myself in November 1978 is still haunting me. It is only partially answered by my research over the past 25-plus years. I offer a few thoughts arising out of my study of Malignant Pied Pipers like the Rev. Jim Jones.

In my opinion, our modern American culture has grown sadly glib, arrogant, and focused on photogenic form, rather than substance of character. TV offers wonderful opportunities for education and artistic discovery. Yet, it is too often focused on the shallow pornography of violence, sensationalism, celebrity gossip, materialism, irresponsible sex, and eroticism. I have watched TV documentaries about Jim Jones, David Koresh, Charlie Manson, and

Marshall Applewhite. These "docudramas" seem to focus more on the exciting and dark charisma of these men, rather than their destructive and severe character disorders. The key question is not addressed: "How might a person recognize those patterns in a leader promptly and perceptively?" And, what can protect a person and his family from a Jones, a Koresh, a Manson, an Asahara, an Applewhite, or a DiMambro/Jouret? It starts with our choice of heroes and our development of an independent thinking, heroic self.

Our Heroes and Antiheroes 'R' Us

Our professional athletes, politicians and even clergy have become primarily entertainers and celebrities. If the 19th century was the century of anxiety and the 20th century, the century of narcissism; then our 21st century is rapidly becoming the century of arrogance, celebrity worship, and hypocrisy. Jim Jones, David Koresh, Shoko Asahara, Marshall Applewhite, Charlie Manson, Joseph DiMambro, Luc Jouret, and Osama bin Laden are out at the extreme end of arrogance. But the passive narcissistic participation of their followers helps put them out there. TV still inadvertently promotes the celebrity status of Malignant Pied Pipers. TV and our educational system seem to promote too much passivity and vicarious participation with charismatic authorities. In essence, beware of all gurus.

Our Media, Our Message, Is Us

Our exciting media trump our inner spiritual substance. The TV programs we watch give evidence of this social psychological fact. TV network programming reflects what our national character is becoming. We, the viewers, ultimately cast our votes with the remote. If we are truthful with ourselves, most of us are "in-betweeners" spiritually, during many stages of our lives. Adolescents in particular are searching for lively affiliation. In fact, we all desire to be a part of a group or cause larger, more noble, and important than ourselves. This desire is psychologically normal. A healthy and effective leader or teacher/mentor is very important in helping our search for meaning and meaningful affiliation.

For example, at Abu Ghraib in Iraq, photographs of torture are the symptomatic evidence of the subtle but powerful connection between the charismatic "Axis of Evil" judgments of our leaders, and the not-so-noble deeds of those they lead. General Taguba's words about Abu Ghraib prison ring in our ears, "A total failure in leadership." Not just sergeants and lieutenants, but commanders-in-chief of preemptive war, and not-so-holy wars.

If the president is on "Meet The Press," the dominant question seems to be, "How did he do?" meaning, "How was his performance?" Did he "act" sincere? Rather than, *was* he sincere, truthful, and clear about his actions, decisions, and policies for our country? Is the American character so weak and phobic that we could not stand the truth from our leader? Wouldn't it be a breath of spiritual fresh air to hear a leader say, "Yes, I made mistakes and underestimated the complexity in post-war Iraq. But I am learning from those mistakes, and we are changing policies in more effective directions and in the following ways...."

In World War II, patriotic Americans were called to sacrifice by our leaders. People saved metal objects and newspapers for the war effort. People accepted gas rationing and reduced driving. Currently, the only sacrifice is our young soldiers' lives and we don't even provide armor for their bodies and trucks. We are asked to go out to shopping malls as usual to restore the economy. We are not asked to participate in helping law enforcement to guard power plants, refineries, reservoirs, train tracks, border fences, or ports. We are pampered with tax cuts. Our leaders' charisma will protect us.

Now let's turn back to the malignant leadership of emperor Jim Jones. It should have been obvious to his followers early on but tragically it was not.

Rev. Jim Jones as Leader and Performer

Listen to Reston's description of a Jim Jones performance, and then Jones's swaggering, forked-tongued words to his flock:

"Occasionally, Jones's religiosity and his Communism got confused in the execution. One night at a People's Rally, he was berating the assemblage for their complaints about the food, scoffing at the way they 'bled' him with their notes about their petty problems, causing him pain in his mind and body. The needs of his followers became an affront to his own infinite, unmatchable Goodness, which in the socialist dominion of Guyana had replaced his former notion of Godliness."

Jones ranted: "People don't like goodness, it makes *them* have to be good. If you see goodness in someone else, it means only one simple equation: you've got to be good too. Folk don't want to be good by nature. Their animal instincts are opposed to being good. They want to be like a goddamn bunch of fighting wolves in a pack. They want to dominate, to kill and not be killed. They want to rob." (Reston, page 180.)

This soliloquy by Jones in search of his parent/himself is pure projection of his inner, stored-up narcissistic wounds and rage. It reveals his inner cynicism, defensive grandiosity, and defective moral leadership. Such free-associative verbal flights by a cult leader often reveal group danger years before the actual apocalypse. Statements like these by a leader must be recognized and challenged promptly and vigorously.

Jones was partially correct however, when he described our human instincts as biologically based. To paraphrase Freud, biology is partially our destiny. However, the evolution of human consciousness has slowly allowed for accumulated moral, ethical, theological, and artistic achievements. Moral leadership takes hard work. It begins with paying the piper of parenthood, as I discussed early in this book. Society's leaders have profound responsibility to purvey genuine moral, ethical, and spiritual values with creativity, dignity, and a good sense of humor. Our children are listening to our leaders and to us.

The "Good Enough" Father

All of the MPPs I have studied have had poor relationships with their fathers. I have tried to think of some of the basic experiences that a relatively healthy relationship with a father provides for a boy and young man. In psychoanalytic terms, a male needs a father figure to identify with at both conscious and unconscious levels. The development of a man's healthy sexual identity and gender-specific behavior are greatly facilitated by a father figure's presence. Social, vocational, marital, and parenting behaviors are at first imitated by a boy and eventually absorbed and consolidated in his adult years.

If the relationship between a boy and his father is solid, the young man will more readily learn how to develop a healthy relationship to authority and authority figures. The inner confidence in his own inner authority allows a young man to be assertive but not destructive and exploitive. He can learn from a father how to treat a wife, a mother, or a sister with loving, caring, and tender respectfulness. The developing man will see how a mature man can disagree with a woman and argue with a man or woman, without losing control, bullying, or becoming cruel, violent, or selfish. This is a precious payment to the piper of parenthood.

The Importance of the Cultural Meaning of Words

We throw around the word "character," but seem to grasp very little about its meaning or content. Four important words to consider in this regard are:

1. Cognition: The act or process of knowing, based on perception, memory, judgment and reasoning,

2. Conation: the aspect of mental life having to do with purposive behavior, including desiring, resolving, and striving.

3. Intuition: direct perception of truth, fact, etc., independent of any reasoning process: immediate apprehension.

4. Character: a word that has 24 domains of meaning in my dictionary! Character is 1. the aggregate of features and

traits that form the individual nature of a person or thing; 2. one such feature or trait; characteristic; 3. moral or ethical quality; 4. qualities of honesty, fortitude, etc; integrity; 5. reputation (good reputation), as in a stain on one's character; 6. distinctive, often interesting qualities; 7. a person with reference to behavior or personality; as a suspicious character; 8. odd, eccentric; 9. a person represented in a drama; 10. a role, as in a play or film; 11. status or capacity; 12. a symbol used in a system of writing; 13. a significant visual mark or symbol; 14. an account of a person's qualities, abilities; 15. a sketch of a particular virtue or vice; 16. any trait, function, structure of an organism resulting from the effect of genes; 17. encoded computer usable data; 18. a cipher; 19. of a theatrical role requiring or having eccentric or comedic, ethnic or other distinctive traits; 20. an actor specializing in such roles; 21. to portray or describe; 22. to engrave or inscribe; 23. in or out of___; 24. In accord with or in discord with one's usual disposition or behavior.

Our American perception, experience of, and collective understanding of the word "character" seems to emphasize meanings 6 and up rather than 1 through 5.

Character of Leader and Follower

Our presidents and politicians make lofty speeches (as grand actors or characters), but they "spin" words and descriptions of events. President Clinton even chose to violate sexual moral boundaries and his responsibility as a role model. He looked us in the TV eye and lied to us. He was impeached, but our elected officials lacked the moral leadership to firmly convict him, dismiss him from office, or formally reprimand him for his actions in front of the nation. What a statement of true leadership that would have been!

President Bush said he was taking the Iraqi war decision to the United Nations. However, when he went to the UN, he told them how it was, and did not listen in-depth to other countries' ideas. Between the lines of Bush's speech, most people could intuit the real message: "I have decided to go to war with Iraq. I want the rubber stamp of UN approval. If I don't get what I want from you, I will go it alone and take my country to war. I am strong and decisive. You are too weak to stand up to the 'Axis of Evil.'" I think many of us had the correct intuition about what he was really saying. He frightened us with the shadowy weapons of mass destruction. At first, Bush never spoke about Saddam's murders of his own people as a major reason to end his rule. Bush never spoke about our former "friendship" with Saddam and how we had supplied him with weaponry.

Overcoming I.Q. Worship
In my opinion, we in America tend to worship or unduly focus on I.Q. and cognition and their application to clever oratory. Our educational process teaches young people math skills, scientific methodology and facts, English and writing skills, and historical facts. But very little study or educational experience is focused on the mature and wise use of intuition, integrity of ethical judgments, and empathy for others in moral decision-making.

Learning to cultivate our emotional and moral I.Q. is difficult, but valuable. The "sixth sense" (intuition) is not easily learned didactically. Our educational system has grown so fearful of being politically incorrect, legally liable and *judgmental*, that our young people don't learn how to make clear judgments.

Understanding intuitively what leaders mean when they talk is difficult, but important for survival. Our celebrities and role models seem to live above the law. They sometimes seem to be vicariously admired for getting away with fraud, rape, and even murder. It is as if a person's morality, compassion, kindness, humility, and empathy are really *not* admired in our modern, "dog-eat-dog" world.

Actually, goodness, moral integrity, honesty, and kindness are "cool" and valuable! Courses that teach ethics, the varieties of moral values, the cultivation of intuition, the meaning of loyalty and true friendship, and independent thinking should start in preschool and be life-long core values of our continuing education. Assessing these areas in one's social, financial, marital, spiritual, religious, and vocational life is crucial for success. The possibility of contentment or happiness in life, or even survival, hinges on the capacity to intuit the difference between following a destructive leader like Jim Jones or a constructive leader like Elie Weizel.

Beware of Psychotherapy Cults
Complexity and stress are ubiquitous in modern life. Various forms of psychotherapy offer to help urban, suburban, and rural citizens in handling stress and anxiety. However, some psychotherapists cross over the boundary of helping their clients. They use dependency-promoting methods of indoctrination and mind control. Any effective psychotherapist accepts some dependency and idealization/pedestalization of himself early in psychotherapy. As successful psychotherapy proceeds towards graduation for the client, however, the therapist is seen as a fallible, mistake-making ordinary human being.

Psychotherapy cult leaders, however, promote dependency and idealization of themselves in perpetuity. They exploit the attachment to themselves for their own power and financial needs. These patterns so typical of exploitive cults are concisely discussed and summarized by Theodore L. Dorpat in his excellent 1996 book called, *Gaslighting, The Double Whammy, Interrogation, and Other Methods of Covert Control in Psychotherapy and Psychoanalysis.* In my opinion, Dorpat's book should be required reading in every psychotherapy training program.

In Chapter 9, Dorpat specifically discusses psychotherapy cults. He points out (page 184) how psychotherapy cult leaders often use shame and humiliation, among other thought-reform techniques (known as "gaslighting"), to induce dependency and self-doubt in their followers. Such mind control leads to destructive and long-lasting psychopathological effects. Dorpat offers six major and cogent characteristics of all destructive cults (page 185). I agree with Dorpat and think his list is a good summary of red flags to alert us to the presence of a Malignant Pied Piper:

175

1. Cult leaders are charismatic, authoritarian, and dominating individuals.

2. Followers join the cult when they are emotionally disturbed and/ or are in transition between developmental states; i.e., when identity needs and security are the greatest.

3. The followers idealize the cult leaders. Both the leader and followers consider the leader to be the supreme authority.

4. The cult leaders suppress the followers' disagreement or opposition.

5. Followers become totally involved in the cult, which often controls every aspect of their personal life, including sex, social relationships, diet, dress, work, and the like.

6. Cults tend to have long-lasting traumatic and destructive psychological effects on followers, who gradually lose their autonomy and their capacity for critical thinking.

I would add that the victims of MPPs also suffer from major perturbations and immaturity in their spirituality.

Final Observations

Not surprisingly, we Americans seem more spiritually hungry than ever. We have few effective contemporary spiritual leaders among us; we read daily about painful, bewildering, predatory pedophile behaviors among our clergy. Physicians, psychologists, and psychotherapists violate ethical and professional boundaries with their "customers." Many lawyers and most politicians worship at the altar of our moneyocracy. People even submit to the ministrations of charismatic psychologist gurus on national TV! In this sense,

Americans are all vulnerable "in-betweeners" at some period in our lives.

Someone like Jim Jones, David Koresh, Osama bin Laden, or even Charlie Manson can capture the rebellious spiritual imagination of neglected young people. An enthusiastic, ambitious, and idealistic young man like my college acquaintance Tim Stoen can see the white and black garb and hear the mesmerizing tune of Jim Jones's utopian pied piper music. One theft doesn't make a thief, one "white night" sermon doesn't make a group revolutionary suicide; but days, weeks, months, and years of listening to the piper's tune can lead our "inner child" away to the mountain of apocalypse.

Only carefully cultivated independence of reason, highly developed intuition, and strength of character can help us recognize and stop a Malignant Pied Piper. I hope this book makes a contribution to that effort.

WORKS CONSULTED

Beck, M. "Children of the Cult." *Newsweek*, May 17, 1993. P. 48-51.

Becker, E. (1973). *The Denial Of Death*. Free Press of Macmillan Publishing, Inc. New York.

Becker, E. (1975). *Escape From Evil*. Free Press of Macmillan Publishing, Inc. New York.

Bedat, A. Bouleau, G. Nicolas, B. (1996). "Les Chevaliers De La Mort." Paris: Tfi Editions. Chapter 3, P. 47-71.

Bergen, P. (2002). *Holy War, Inc*. Touchstone of Simon & Schuster, New York.

Bodansky, Y. (1999). *Bin Laden: The Man Who Declared War On America*. Forum of Prima Publishing, Crown Publishing, Random House, New York.

Breault, M. and King, M. (1993). *Inside The Cult: A Member's Chilling, Exclusive Account of Madness and Depravity in David Koresh's Compound*. Signet Books, Penguin Publishing, New York.

Browning, R. and Greenway, K. (1997). *The Pied Piper of Hamelin*. Dover Publications Inc., Minneola, New York. [Original Publisher (1888) George Routledge and Sons , London.]

Bugliosi, V. and Gentry, C. *Helter Skelter*. 1974 W.W. Norton and 1975 Bantam Paperback.

Campbell, R. and Edgerton, J. (1994). *American Psychiatric Glossary*, 7th Edition. Appi Press, Washington, D.C.

Clark, J. (1993). "On the Further Study of Destructive Cultism." *Psychodynamic Perspectives On Religion Sect And Cult*. Ed., David Halperin. Publisher, John Wright-Psg, Inc. Boston, Bristol, London. Chapter 24.

Dennis, A. (2002). *Osama Bin Laden: A Psychological and Political Portrait.* Wyndam Hall Press: Lima, Ohio.

Deutsch, Alexander (1980). "Tenacity of Attachment to a Cult Leader." *American Journal of Psychiatry.* Vol. #137, #12. P. 1569-1573.

Dorpat, T. (1996). *Gaslighting, The Double Whammy, Interrogation, and Other Methods of Covert Control in Psychotherapy and Psychoanalysis.* Jason Aronson Press, Northvale New Jersey, London.

Erikson, E. (1950). *Childhood and Society.* W. W. Norton Publishing, New York.

——— (1969). *Gandhi's Truth: On the Origins of Militant Non-Violence.* W. W. Norton Publishing, New York.

Fields, R. (1976). *Society Under Siege: A Psychology of Northern Ireland.* Temple University Press, Philadelphia.

——— (1986). "The Psychological Profile of the Terrorist." Paper presented at the American Psychological Association Convention, Washington, D.C.

Fornari, F. (1966). *The Psychoanalysis of War.* University of Indiana Press. Bloomington, Indiana.

Freud, A. (1936). *The Ego and the Mechanisms of Defence.* International Universities Press, New York.

Freud, S. (1900). "The Preface to the Second Edition of the Interpretation Of Dreams," Vol 4. *The Standard Edition of the Complete Works of Sigmund Freud.* Translator/Editor, James Strachey. The Hogarth Press, London. First Edition, 1957. Pxxvi.

——— (1914). "On Narcissism: An Introduction," in *The Standard Edition,* Vol. 14, P. 67-107.

——— (1916). "Some Character Types Met with in Psychoanalytic Work," in *The Standard Edition,* Vol. 14, P. 309-333.

——— (1921). "Group Psychology And The Analysis Of The Ego," in *The Standard Edition,* Vol. 18, P. 63-143.

——— (1919) ."The Uncanny," in *The Standard Edition.* Vol. 17, P. 217-256.

Friedman, T. (2002). *Longitudes and Attitudes.* Farrar, Straus, Giroux, New York.

Gleeick, E. and Lacaayo, R. "Inside the Web of Death," *Time*: Special Report, April 7, 1997, P. 45-131.

Goldberg, C. (1996). *Speaking with the Devil: A Dialogue with Evil.* Viking Penguin, New York.

Grinberg, L., Sor, D., and Tabak De Blanchedi, E. (1997). *Introduction to the Work of Bion.* Jason Aronson Inc. New York.

Gunaratna, R. (2002). *Inside Al Qaeda: Global Network of Terror.* Berkley Books, New York.

Halperin, D., Editor, *Psychodynamic Perspectives on Religion, Sect, and Cult.* John Wright Pub. Inc., 1983. Boston, Bristol, London.

Hofer, E. (1951). *The True Believer.* Harper Brothers, New York.

Janus, I. and Mann, I. (1972). *Victims of Groupthink: A Psychological Study of Foreign Policy Fiascoes.* Houghton Mifflin, Boston.

Johnson, A. and Szurek, S. (1952). "The Genesis of Antisocial Acting Out in Children and Adults." *The Psychoanalytic Quarterly,* Vol. 21, P. 323-342.

Kernberg, O. (1986). *Severe Personality Disorders: Psychotherapeutic Strategies.* Yale University Press, New Haven, London.

Francis, A. (1994). *DSM-IV, The Diagnostic and Statistical Manual of Mental Disorders, Edition IV,* of the American Psychiatric Association. Appi Press, Washington, D.C. P. 658-661 ff.

Kohut, H. (1971). *The Analysis of the Self.* International Universities Press, New York. P. 64-67.

———— (1978). "Creativeness, Charisma, Group Psychology," in *The Search for the Self: Selected Writings of Heinz Kohut: 1950-1978.* Vol. 2, Ed. by Paul Ornstein. IUP, New York.

Langone, M. (1993). *Recovery from Cults.* W.W. Norton, New York.

———— (1996). Article in Cults in *Psychiatric Times,* July, 1996. P. 14.

Layton, D. (1998). *Seductive Poison: A Jonestown Survivor's Story of Life and Death in the People's Temple.* Anchor Books, New York.

Lifton, R. (1989). *Thought Reform and the Psychology of Totalism: A Study of "Brainwashing" in China.* The University of North Carolina Press, Chapel Hill and London.

Maslow, A. (1954). *Motivation and Personality.* Harper Books, New York.

Mayer, Jean-Francois. "The Solar Temple," *Nova Religio* Vol. 2, #2, April 1999. Also, personal communication with the author about psycho-biographical information on Luc Jouret and Joseph DiMambro. March 13, 2001.

Nichols, W.R., Editorial Director. *Random House Webster's College Dictionary* (2001). Random House, New York. P. 1002, "Pied Piper."

Olsson, P. (1980). "Adolescent Involvement with the Supernatural and Cults." *The Annual of Psychoanalysis,* Vol. VIII. P. 171-196.

Osama Bin Laden and Dr. Abd-Ar-Rahman Al-Ashmawi. Poem, "The Travail of a Child Who Has Left the Land of the Holy Shrines." *The New York Times,* April 2, 2002, P. 16.

Palmer, S. "Purity and Danger in the Solar Temple." In *Journal of Contemporary Religion* (London), Vol. 11 #3, October 1996. P. 303-318.

Reiterman, T. with Jacobs, J. (1982). *Raven: The Untold Story of the Rev. Jim Jones and His People.* E.P. Dutton, Inc. New York.

Reston, James Jr. (1981). *Our Father Who Art in Hell: The Life and Death of Jim Jones.* Times Books, New York.

Riley, M., Woodbury, R., Johnson, J. and Shannon, E. *Time* Magazine: Special Report on Waco. May 3, 1993. P. 35-43.

Rimer, S. and Verhoven, S.H. "Youngsters Tell of Growing Up Under Koresh," *The New York Times,* May 4, 1993, National Section.

Robinson, A. (2001). *Bin Laden: Behind the Mask of the Terrorist.* Arcade Publishing, New York.

Sayle, M. "Letter from Tokyo: Nerve Gas and the Four Noble Truths." *The New Yorker,* April 1, 1996.

The Book (Bible), (1985). Tyndale House Publishers. Wheaton, Illinois.

Thibodeau, D. with Whiteson, L. (1999). *A Place Called Waco: A Survivor's Story.* Public Affairs Press, New York.

Said, E. (1986). *After the Last Sky: Palestinian Lives.* Pantheon Books of Random House, New York.

Singer, M. (1995). *Cults in Our Midst: The Hidden Menace in Our Everyday Lives.* Jossey Bass Press, San Francisco.

Stein, M. (1991). "Dreams, Conscience And Memory." *Psychoanalytic Quarterly* Vol LX, P. 199.

Stern, J. (2003). *Terror in the Name of God: Why Religious Militants Kill.* Harper Collins, New York.

Sullivan, H.S. (1940). *Conceptions of Modern Psychiatry.* W.W. Norton & Co., New York, P. 16.

Van Biema, David. "Prophet of Poison." *Time* Magazine, April 3, 1995. P. 27-41.

Volkan, V. and Itzkovitz, N. (1984). *The Immortal Ataturk: A Psychobiography.* University of Chicago Press.

――― (1988). *The Need to Have Enemies and Allies: From Clinical Practice to International Relationships.* Jason Aronson Inc, Northvale, New Jersey, and London.

――― (2004). *Blind Trust: Large Groups and Their Leaders in Times of Crisis and Terror.* Pitchstone Press, Charlottesville, Va.

Wilson, C. (2000). *Rogue Messiahs: Tales of Self-Proclaimed Saviors.* Hampton Roads Press, Charlottesville, Va.

Wright, L. (1993). "Orphans of Jonestown." *The New Yorker,* Nov. 22, 1993, P. 66-89.

INDEX

Abrams family, Ramtha cult, 48–50
"Cindy," 121–25
Stoen family, People's Temple, 48–50
"Tom" and "Tanya," 113–19
castration, 94, 100, 101
cause vs. effect argument, 15
Center for the Preparation of the New Age, 106
channeling, 35, 97
character, meaning of, 172–73
charisma of cult leaders, 71, 79, 100, 107
chemical science vs. psychiatric science, 24
child-admirers of cult leaders, 37, 154
Childhood & Society (Erikson), 40
childhood trauma/neglect
    bin Laden, 136–39
    general, 18, 38, 161–62
    Jones, 45
    Koresh, 65, 72–74
    and later child exploitation, 64–65, 68, 75–76
    Manson, 81
    terrorism and children, 156
Chizuo (Asahara), 86
    See also Asahara, Shoko
Christfigure delusion
    Asahara, 88–89
    Jones, 53, 55
    Koresh, 71, 74
    Manson, 79, 85
Christian influences on cult leaders
    Applewhite, 97
    Jones, 43–44
    Koresh, 64, 66, 69–71, 74–76
    Manson, 81
Clark, John G., 15
Clayton, Stanley (People's Temple member), 49

Printed in the United States
70075LV00002B/72

9 781413 776683